HEIFER INTERNATIONAL is a nonprofit, humanitarian organization dedicated to ending world hunger and caring for the earth. Heifer pursues this mission by providing livestock, trees, training and other resources to help struggling families move toward greater self-reliance and build sustainable futures. Heifer's gift of "living loans" offers milk, eggs, meat, wool, draft power and other benefits that become improved nutrition, health, education and income for resource-poor families.

Heifer partners with groups to create a development plan with specific goals based on the values of their community. Partners learn to care for animals and grow crops in ways that can be sustained for future generations. Heifer adds expertise in animal health and husbandry, water quality, gender equity, agroecology and community development.

Over the years, Heifer has developed a set of guiding principles called The Cornerstones for Just and Sustainable Development. The Cornerstones form the acronym "PASSING GIFTS," an essential element of our sustainable approach. Heifer requires that livestock recipients "pass on the gift" of one or more of their animal's offspring and training in environmentally sound agriculture. In this manner, an endless cycle of transformation is set in motion as recipients become equal partners in ending poverty and hunger. Since 1944, this common sense approach to sustainable development has enabled Heifer to partner with millions of families in more than 125 countries to improve their quality of life.

TABLE OF CONTENTS

Chapter 3—Nutrition and Feeding

Chapter 4—Forage Production, Pastures and Environmental Management

Chapter 5—Reproduction

Chapter 6—Kidding

Chapter 7—Milking the Doe

Chapter 8—Health Care For Your Goats

Chapter 9—People, Economics and Marketing

Chapter 10—Humane Transportation and Slaughter of Goats

HEIFER INTERNATIONAL'S CORNERSTONES FOR JUST AND SUSTAINABLE DEVELOPMENT

PASSING ON THE GIFT

"Passing on the gift" is the heart of Heifer International's sustainable community development philosophy. Every family who receives an animal signs a contract to pass on one or more of their animal's offspring to another family in need, along with the training and skills that they have acquired. This unique approach creates a ripple effect that transforms lives and communities.

ACCOUNTABILITY

Heifer provides guidelines for planning projects, screening recipients, monitoring progress and conducting self-evaluations. The groups define their own needs, set goals and plan appropriate strategies to achieve them. They are also responsible for submitting semi-annual monitoring reports to Heifer International.

SHARING AND CARING

Heifer believes that global problems can be solved if all people are committed to sharing what they have and caring about others. One of our most important Cornerstones, sharing and caring is an integral part of our vision for a just world.

SUSTAINABILITY AND SELF-RELIANCE

Because Heifer funds projects for a limited time, project groups must devise strategies for its continuity. In our experience, self-reliance is most easily achieved when a group has varied activities and generates support from several sources.

IMPROVED ANIMAL MANAGEMENT

Feed, water, shelter, reproductive efficiency and health are the essential ingredients in successful livestock management. The animal must be an appropriate breed for the area and should be a vital part of the farm activities without placing an extra burden on the family or resources

NUTRITION AND INCOME

Livestock contribute directly to human nutrition by providing high-quality protein. Indirectly, they provide draft power for cultivation and transportation, as well as manure for soil fertility. Livestock provide income for education, healthcare and housing, and as living savings accounts, provide long-term economic stability.

GENDER AND FAMILY FOCUS

Gender refers to the socially-defined roles of women and men in each culture. Heifer encourages women and men to share in decision-making, animal ownership, labor and benefits. Heifer has a gender initiative, as well as its Women in Livestock Development (WiLD) program.

GENUINE NEED AND JUSTICE

Heifer is a partner to people in need who can improve their quality of life with modest support. Priority is given to marginalized groups. The poorest in the community should be included and receive priority for assistance. Families are eligible regardless of creed or ethnic heritage.

IMPROVING THE ENVIRONMENT

The introduction of Heifer International projects should have a positive impact on one or more of the following: soil erosion, soil fertility, sanitation, forestation, biodiversity, pollution, wildlife and watershed conditions.

FULL PARTICIPATION

Members of the group "own" the project and have control over all key decisions. Heifer is committed to involving all members in decision-making, working with grassroots groups to develop strong leadership and organization.

TRAINING AND EDUCATION

Groups determine their own training needs, and local people are involved as trainers. Training includes formal sessions as well as informal farm visits and demonstrations. In addition to training in livestock husbandry and environmental conservation, groups have requested training in food processing, marketing and human nutrition.

SPIRITUALITY

Spirituality is common to all people, regardless of their religion or beliefs. It is expressed in their values, sense of connection to the earth and shared vision of the future. Often spirituality creates a strong bond among group members, giving them faith, hope and a sense of responsibility to work together for a better future.

PREFACE

To the reader:

Heifer International is pleased to present this new, revised edition of *Raising Goats for Milk and Meat*. Since the first edition was printed in 1984 literally hundreds of good texts, teaching materials and books on goat husbandry have emerged. Our goal in 1984 was and continues to be publishing a comprehensive manual for the beginning goat farmer. To this end, the 2008 edition features Learning Guides at the beginning of each chapter, an updated health section and additional information on marketing, raising meat goats and caring for the land. This version also includes stories illustrating how goats have contributed to the livelihoods of small scale farmers around the world.

I am fortunate to have worked for Heifer International for more than 40 years, serving in many capacities including regional director, director of development and major gifts officer. During the 1970's and 80's, as a regional director I was responsible for procuring and shipping animals to Heifer programs around the world. In those "olden days," I handled procurement, daily care of the animals waiting for shipment, preparation for shipping, health tests and arrangements for air transportation. Back then, goats far out-numbered the other species that were shipped and I took an avid interest in them.

This keen interest led to an invaluable hands-on experience during my sabbatical in 1979 when I worked at La Fiesta Farm in Paso Robles, California, owned by Judy Patrick. At the time, tatiana Stanton* was the herdsperson and I had the opportunity to work with 75 milking does and numerous kids. In 1980 and 1981, I taught courses in dairy goat husbandry in Cameroon, West Africa. When I arrived, I found limited teaching materials with basic information and simple suggestions about goat husbandry. I then began to assemble a teaching guide for the sessions in Cameroon, which evolved into a small training manual. When I began a Masters of Animal Science program at the University of Connecticut, *Raising Goats for Milk and Meat* became my Master's project. From 1984 through 1989, I taught goat husbandry courses at the University of Connecticut and extended my goat activities by becoming secretary-treasurer for the International Goat Association from 1993-2000.

My co-author, **Paul Rudenberg, DVM,** has served as a volunteer for Heifer International since 1983. As the Heifer International Country Director for Haiti from 1999 until 2005, he helped initiate projects focusing on small ruminant and cattle husbandry, soil conservation, reforestation, literacy and participatory training. Dr. Rudenberg is a graduate of Middlebury College and Cornell University School of Veterinary Medicine. He has lived in Haiti for 17 years where he has also worked with the Inter-American Institute for Cooperation on Agriculture (IICA), Church World Service, Christian Veterinary Mission, Oxfam-UK, and the SEED Agricultural School. Paul and his family live in Merci, Haiti.

Barbara Carter started drawing when she was a young child. The daughter of Shirley Carter, a well-known water color artist and teacher, Barbara depicted animals in art from an early age. In the 1970's, Carter lived and worked on a goat farm owned by her sister Anne Bossi, who for 20 years was the field person for Heifer International's programs in the Northeast United States. Carter lives on a small farm in Randolph, Vermont and continues her childhood passion as a commercial artist. She is also the illustrator for Heifer International's publication: *The Heifer Model: Cornerstones, Values Based Development*.

Rosalee Sinn

ACKNOWLEDGEMENTS

We want to acknowledge the core group of animal husbandry specialists who contributed extensively to the writing, editing and reviewing of this book.

Terry Wollen, DVM, is the Director of Animal Well-Being and Staff Veterinarian with Heifer International, working with Heifer staff and partners around the world to raise the level of excellence in animal health and well-being and seeking grants to further training in improved animal management. Dr. Wollen's background is in food animal medicine and production. Following two years in the Army Veterinary Corps, he worked in a four-man dairy, beef and equine veterinary practice in the state of Idaho, followed by 20 years in research and development with the Bayer Corporation, Animal Health Division. Committed to international development, he worked four years with the United Methodist Committee on Relief in Armenia and three years with Heifer International in Asia before joining Heifer International in Little Rock, Arkansas in 2004.

Beth A. Miller, DVM, is President of Miller Agricultural Consulting and an Instructor of Anatomy and Physiology at Pulaski Technical College in North Little Rock, Arkansas. Her expertise includes animal health, with a specialization in small ruminants and sustainable grassroots development. She served as Director of Heifer International's Gender Equity Program for 10 years, and spent eight years in private practice. She was secretary-treasurer of the International Goat Association from 2000-2002. In the 1980's, Dr. Miller and Paul Yoder, her husband, operated a small-scale goat dairy called Bloom Dairy in Baton Rouge, Louisiana, which produced fluid milk, cheeses and ice cream. Dr. Miller is a graduate of Louisiana State University School of Veterinary Medicine.

An Peischel, PhD, is the founder of Goats Unlimited, Ashland City, Tennessee. She has been utilizing meat goats for land and stream bank enhancement, procurement, fuel load reduction and riparian (water bank) restoration for the past 22 years. Goats Unlimited, begun in 1985, is a Kiko goat breeding center and farm that practices holistic management, pasture/browse management and sustainable alternative agriculture. Dr. Peischel has taught Small Ruminant Production at the University of Hawaii at Hilo, University of California at Chico, and is now the Small Ruminant Extension Specialist for Tennessee State University and the University of Tennessee.

tatiana Luisa Stanton, PhD, is an Extension Goat Specialist at Cornell University. She operates a small pasture-based meat goat herd of 35 breeding does in Trumansburg, New York. From 1976 to 1983, tatiana worked in the Caribbean as a small ruminant specialist through the Peace Corps for Heifer International. She was the herdsperson for La Fiesta Goat Dairy in Paso Robles, California and for the International Dairy Goat Research Center in Prairie View, Texas. Currently, Dr. Stanton conducts workshops on evaluating goats for market readiness, small ruminant integrated parasite management and pasture management for goats. She works with farmers to evaluate the effectiveness of management practices such as out-of-season breeding, creep feeding, flushing and the use of locally obtained whole grains and by-products in goat rations. Dr. Stanton holds a PhD in Animal Breeding from Cornell University and a M.S. in Agriculture from San Luis Obispo University, with emphases in dairy science and tropical soils.

**tatiana does not capitalize the first letter of her name.*

ACKNOWLEDGEMENTS

Ann Wells, DVM graduated from Oklahoma State University School of Veterinary Medicine and has more than 20 years experience in livestock production, including producing and selling lamb and grass-fed beef. In private practice for 11 years and having worked for a sustainable agriculture organization for nine years, Dr. Wells now has her own business, Springpond Holistic Animal Health, in Prairie Grove, Arkansas, where she develops sustainable animal wellness plans and trains producers and educators. She has been involved with organic livestock production on her own farm and working with other organic producers for 15 years. Dr. Wells currently advises the staff at Heifer Ranch in Perryville, Arkansas, researching parasite management strategies to reduce the need for anthelmintics. She also works with Heifer projects in the U.S. to improve their livestock production by focusing on animal well-being.

Dilip P. Bhandari, BVSc & AH, MVM, joined Heifer International as Program Officer for Animal Well-Being in September 2006. Dr. Bhandari received his BVSc and AH from the Institute of Agriculture and Animal Science, Tribhuven University, Nepal in 1999 and MVM from Seoul National University, Seoul, South Korea in 2006. He has worked with a national NGO, Animal Health Training and Consultancy Service (AHTCS) in Nepal training grassroots para-vet professionals. He is experienced in the field of Community Animal Health Workers (CAHWs) and grassroots training related to livestock health and management. Dr. Bhandari also served as training officer in Heifer International's Nepal program for two years.

Marie McCabe, DVM, joined Heifer International in June 2006 as Director of Community Education. Dr. McCabe earned her veterinary degree from Ohio State University in 1982. She served as Director of the Veterinary Technology Program at Columbus State Community College and as an extension veterinarian at the Center for Animal-Human Relationships at the Virginia-Maryland Regional College of Veterinary Medicine. She is former President of the American Association of Human-Animal Bond Veterinarians and serves on the Council of the International Society for Anthrozoology. Dr. McCabe served as a Veterinary Medical Officer with the Veterinary Medical Assistance Team of the National Disaster Medical System at the World Trade Center disaster.

Special thanks to **David M. Sherman, DVM, MS, Diplomate, ACVIM,** for reviewing the lessons on goat health.

In addition we wish to acknowledge the contributions of the following individuals who have been active in the fields of goat husbandry, sustainable development and training.

Adel M. Aboul Naga
Anne Bossi
C. Devendra
James De Vries
George F. W. Haenlein
Gordon Hatcher,
Joseph Howell
Long Housheng
Gan Jiyun
Erwin Kinsey
Wallace Roby
William H. Rose, III
Ann Starbard
Susan Stewart
Elsa Swenson
Ron Tempest
Ellen Zepp

INTRODUCTION

Goats are some of the most beneficial animals in the world, providing meat, milk, fiber, fertilizer, and draft power in addition to working as partners in land reclamation. Widely known as the "poor man's cow," goats have some under-recognized advantages over other animals. They are readily adaptable, thriving in tropical, cold, dry or humid climates. Given their small stature compared to other livestock, goats can be raised on large or small land holdings. Furthermore, approximately two-thirds of the feed energy used to raise these animals comes from substances which are undesirable, indigestible and inedible by humans.

It is little wonder that goats were among the earliest domesticated animals; records in this regard date back 10,000 years to the Tigris-Euphrates Valley. This may be explained by the fact that goats have gentle temperaments, making them ideal household animals. When the cost of cattle-raising is prohibitive, goats cost very little, are ideal for family milk and meat production and can be easily sold for income. The milk and meat produced by one goat is the perfect balance: it is sufficient to meet children's nutritional requirements, without the storage problems associated with the larger supply produced by cattle. In warm climates where no refrigeration is available, the meat from one goat can be consumed by a family before it spoils. And as there are few religious taboos related to the consumption of goat meat and milk. Goat meat, dairy products or offspring can easily be sold for extra income.

Traditionally raised for milk and meat, goats are the source of the most widely consumed meat in the world. It is an excellent source of protein. It is low in fat and cholesterol and high in vitamins and minerals. Similarly goat milk is more widely consumed worldwide than cow's milk and, for many, is easier to digest. Its rich, complex flavor and nutritional qualities has helped the goat cheese industry become a major niche market in Europe and the United States. Goat products have become sought-after commodities in developed countries.

In 2005, the world's goat population was approximately 800 million, according to the United Nations Food and Agriculture Organization (FAO). At least 80 percent of the world's goats are found in developing countries, which also produce about 75 percent of the world's supply of goat skins. For centuries, goat skins have been used for drums, clothing, parchment and as containers for water and milk. In modern day, goat skins are used to produce high quality leather goods, cashmere and mohair that end up in luxury markets all over the world.

Since their domestication over 10,000 years ago, goats have held a special place in the hearts and cultures of many people. With their many benefits, it is clear that goats will continue to have an important role in sustainable agricultural development and food production around the world.

HOW TO USE THIS BOOK

Although this book will be a useful reference for an individual goat farmer, it is primarily designed as **a tool for a facilitator** to use when training a group of farmers interested in starting or improving goat production. The book is organized into chapters and each chapter has two parts: The Learning Guide and The Lesson. How the training is organized depends on the experience and needs of the group.

The facilitator will want to know as much about the group as possible. If there are interested goat farmers but they are not yet organized into a "goat producers group," forming a group should be a priority. There are many government and NGO resources that can help grassroots groups learn to set objectives, elect officers and organize finances. In fact, excitement about goats is one of the best ways to establish a grassroots group. As members work together and build trust, they can learn from and assist each other.

PARTICIPATORY METHODS

Heifer International has found that *Participatory Training* is the most effective method of introducing new livestock technology to a group of adults. Unlike traditional lecture formats, where the teacher has the knowledge and the students are asked to memorize it, the participatory method puts the adult learner in the driver's seat, and invites the facilitator to be a learner as well. Facilitators work **with** participants in setting objectives and agendas, rather than telling them what to do.

GROUP BACKGROUND

If you are working with an existing group, you will want to know the number, age, income and gender of members, previous activities, current number of goats and husbandry practices, and supply and marketing chains. Find out the level of literacy, and the language preferences of the group.

CHALLENGES, ISSUES, THE THINGS PEOPLE TALK ABOUT

Before planning training be sure visit local farms and listen to conversations between farmers discussing community and livestock issues. When they come together, what do they talk about? How is it related to goat production? Are the prices low for milk or goats? How is that affected by the market or the intermediary? Are there issues of land ownership or decreased productivity of the land? Are there issues of livestock destroying crops? Discover the issues and include them in your training design as places to stimulate discussion and problem-solving about the real challenges the farmers are facing.

REVIEW OF PLANNING FOR A TRAINING SESSION

The seven steps of planning for any good participatory training event are: Who, Why, Where, When, What, What For, and How. (Vella, Jane. *Training Through Dialogue*)

- **Who** is the most important. When livestock training is offered, often the owner of the animals attends, although he may not be the actual daily caretaker. Encourage both owner and caretaker to attend, if not the same person. In developing countries, the male head of household is often the owner of the animals, especially cows, but the women and children in the family or hired help do the actual work. In the case of goats, women often receive the income from milk or sale of kids. It is always more effective to train the person who works with the animals, so information is not lost, and they can ask relevant questions. Therefore, it is important that the women, children or workers are invited, and the owners are supportive. This may take some preliminary work before the technical training begins.

We begin our planning with who will attend because the people who attend should set the agenda for the training. What do the participants want to learn about? What are the issues the participants are dealing with in raising livestock, in their village and region? What is important to them? Traditional training starts with the teacher and what the teacher wants to teach or thinks the learners should know. Participatory training is learner-centered and starts with what the learners want to know and can apply immediately.

Language, both spoken and written, must be appropriate for the participants. Women and men with little formal education may be fluent in local languages but not the national language. Their experience and expertise with goat raising is very important so their participation is important. Replacing written information with careful drawings or photographs may be more appropriate for participants who do not use the written language or when there are multiple languages in the group. Arrange translation if necessary. Avoid technical jargon when speaking to farmers. **The value of the training is what people go home and do!**

- **Why** do the participants want the training? How will they use it? Why are they investing valuable time to attend the course. This should influence the content of the training. For example, if the participants are coming because they want to treat illnesses in their goats, they will not be happy if the whole training is about nutrition and pastures (even though it is known that good nutrition will prevent disease). What do the participants expect to gain? Are they motivated to change their management, or are they happy with current conditions? Are they required to attend to qualify for a free goat? The more decisions the group makes about the purpose, place and content of the training, the more successful it will be.

- **Where** means where the training is held. In general, it is best to use group members' farms, so the lessons can be immediately applied to an actual situation. The farm does not need to be a "model,' in fact, the more needs, the better! The group can analyze needed improvements, and even help the owner make them. If an existing community building is more convenient due to a central location, be sure to arrange hands-on activities with live animals, actual feedstuffs, etc. If there will not be enough goats for everyone to do hands on practice, consider finding a different location where there are adequate goats for practice sessions. Be sure the location is convenient for the farmers as well as the facilitators. Some farmers may live some distance from the training so it will be important to arrange for transportation. The agenda should include a clear order of events and plenty of time to see the farm and address questions. Locate toilets and clarify plans for meals at the beginning of each session.

- **When** means the time of year or season, the time of day for the training and the amount of time required. The time should be convenient for the farmers and not just the facilitator! Also, men and women may be busy at different times of day, so separate training times may be necessary. Farming has seasonal workloads, so training may need to stop during the planting or harvest season, and pick up again during the "slow" season. Often the rainy season is the time when farmers have the most time yet access to villages may be difficult. Careful planning to allow the training to happen when the farmers are free and goats are available is very important.

  Most of these lessons may be adapted for different time frames, from two to four hours. If a residential training course is offered, find out if the men and women in the group prefer two days, three days or two weeks. Cost of transport, food and housing needs to be arranged well in advance.

- **What** is the topic of the lesson, such as goat nutrition or reproduction. Choose topics that are relevant and can be applied right away by farmers. For example, don't discuss planting pastures and trees during the rainy season when it is too late to plant or the need for a well when it will be months until it is dry enough to dig one. Choose topics that are central and important to know, not just nice to know. Find out current levels of knowledge and practice, either before the session or at the start. If there are different levels of experience, you may want to divide the group, or design sessions so that more experienced farmers have ample opportunity to share in very practical ways. Be sure to adapt the lesson to the abilities of the group, both economic and educational. Always start where the people are, not where you think they should be. If they complete a simplified training, and slightly improve their production, you can always offer more advanced training in the future.

- **What for** refers to the learning objectives. It is very helpful in planning to state clearly what the participants will actually DO by the end of a session. This helps facilitators to refrain from saying participants will "learn, understand, know" a material. How do we know that we know something? Only when we do it! So, state what the learners will actually do in a session. For example: "By the end of the session, participants will have designed and built a model goat pen." If the learners will not do anything active in your session you can ask yourself how you can change your design to make it active and participatory. This clarifies the planning process and is very helpful. At first it is challenging to do this, but in the end it will help increase the benefits of the training. Learners should know at the beginning of a workshop or session what they will do during the session. That way they can decide how participative they want to be and how useful the session will be for them.

- **How** means what is your actual lesson plan, and what materials do you need to prepare in advance. Your lesson plans should list specific times for activities or discussions, but be flexible. Don't lose an opportunity for important insights by the group in order to stick to a preordered list. Gather markers and flip charts for writing lists, or making drawings, well in advance. Start a discussion about challenges and issues or current practices: whether livestock or fire are damaging crops or the challenges of feeding, breeding or marketing. Drama can be a fun and effective way to learn. Bring needed props so small groups can act out typical situations, and lead a discussion about different solutions. Be prepared to intervene if certain dominant individuals try to do everything, or if the men "help" the women and prevent them from learning to give an injection or castrate a kid. Think about how to organize small groups for practice and discussion sessions so that women or young people are given a voice and men do not dominate all. Discover the local language for body parts, diseases, and local medicines and use them. In the case of diseases that can be prevented by vaccination, discover the local name and teach the appropriate name to enable the farmer to buy the correct vaccine.

Make sure you plan plenty of time in your sessions for the learners to reflect on what they have learned and practiced. For example, if they practice giving injections, schedule a time at the end of the practice not only to review what they have learned but also to reflect on it. What was surprising to them? What was useful? What do they know that is different than what they knew before? What can they do as soon as they get home? What will be hard for them in applying this at home and how will they overcome the difficulties? These kinds of questions for group reflection allow an individual to solidify their practical experience and think about applying it in their own animals. This will create a very powerful learning experience, more powerful than just practice alone.

Start a training session by asking for their expectations and current beliefs or practices. It is good to review highlights from earlier sessions, and show how they relate to the current topic. Try to use less theory and scientific terminology and more practical advice such as the amount of concentrate to feed a doe per liter of milk produced. Still, remember to challenge farmers to consider the big picture about how they want their lives to be, and the practical steps that can bring about change.

Raising Goats for Milk and Meat provides an experienced-based training model on the basic care and management of dairy and meat goats. Although accurate information is important for successful training, the training process itself is crucial. Individuals learn best when the topic is of immediate practical value and they can share their experiences and skills with others.

The design needs to be culturally appropriate for the learners and consistent with their current resources. Learning groups from different classes, cultures or sexes may determine how we approach the training sessions. The group can help in determining accepted standards of behavior. Be skeptical of a single individual's opinion of what "all rural Maasai believe" or "all women think," especially if they are not part of that group.

The course can also be used by individuals, utilizing the advice and assistance of other goat breeders, extension personnel and local veterinarians.

Remember that this manual is an introductory handbook, and does not replace local expertise, and the more advanced resources listed at the end of the book. The Learning Guide at the beginning of each chapter provides suggestions, but the group and the facilitator should determine the most effective activities for the group.

Goat Production Mitigates Poverty Impact

▲ *Duch Sakhorn (second from left) with his family.*

Starting in early 2005, the Household Goat Production, a project supported by Heifer International Cambodia and the Farmers Goat Association in Prek Ta Ong Village, Peam Ok Nha Ong Commune, Lvea Em district, Kandal Province provided direct support to 16 poor families. They received training in Heifer International's Cornerstone Model and in goat husbandry. The project goats were to assist the family in improved nutrition, marketing goat products and generating income.

Although goats have been raised by Muslim farmers, they were not the traditional livestock species kept for food by the majority of farmers in Cambodia, who kept cows and water buffalo. However, the project has been successful in assisting poor farmers in raising goats for food and to demonstrate the adaptability of goat production for small farms.

Each family received two female breeding goats and a micro-credit loan for setting up a goat shelter. Before receiving goats, the project participants were given a series of training courses on feeding goats, common diseases, goat management, zero grazing and compost fertilizer development. The project promotes the careful ensiling of cassava leaves and fodder trees including kapok and Leucaena. Manila Tamari and Mulberry were grown around their land for leaves as a protein source. To insure goat health, a key farmer in the community was selected for training by a Village Animal Health Worker.

For the past two years, the Farmer Goat Association has demonstrated the significant impacts and direct improvement of family income from sale of goat meat. The original recipient families have "passed on the gift" of offspring to other families in the community while remaining offspring have been sold for cash to support their family needs and pay for their children's education. Parents and children share household work and goat care. Children like the goats very much and help feed them after school time. As mixed crop-

Continue on page xix

Continued from page xviii

▲ *Khim Sonem with her goat.*

and-livestock farmers, the project participants are using goat manure as compost fertilizer to keep soil productive. They grow sweet potatoes, maize, beans and rice.

Mr. Duch Sakhorn, his wife and five children are a poor family living in Prek Taong village. They joined the project and received two breeding goats along with training. The two original placements produced eight goats of which four were passed on to other poor families in the community. Last year, Mr. Sokhorn's family sold one goat for cash to support their five children for school; "I have no problem in keeping the goats," said Mr. Duch Sakhorn, adding, "My children help take care of them. We also make compost fertilizer from goat manure for fertilizing our one-half hectare field."

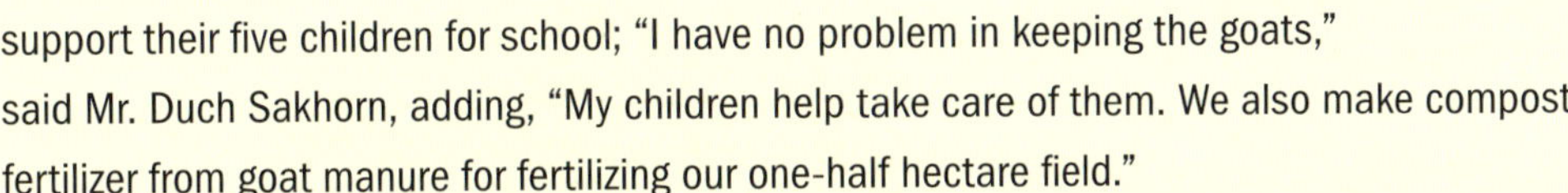

When asked which of the 12 Cornerstones he thought was the most important in his family livelihood, Mr. Duch Sakhorn said: "I am inspired by the Cornerstone "Gender and Family Focus" because my family members always share household tasks and receive benefits equally."

Mrs. Khim Sonem, a poor single mother with two children, is another beneficiary. Her family earns their living by growing banana trees, beans, and maize. She and her children have three goats, 13 chickens and two pigs. She saves goat and pig manure for fertilizing crops saving her the expense of purchasing chemical fertilizer. Since participating in the project, she has sold three goats for other supplies needed for her family. "My family's living condition has improved as we get additional income from selling goats," said Mrs. Kim Sonem. She hopes to add two more does to her herd next year.

Although goat production in the community is progressing, the market for goat products is limited. Goat meat and milk are not common at local markets in Cambodia, yet the Farmer Goat Association is sharing ways that goats can help poor families with limited farm land. They encourage the use of goat meat to improve family nutrition, which is low in protein. The Farmer Goat Association of Lvea Em Commune works together to introduce the use of goat meat. The association offers dishes of goat meat for ceremonial or celebration events in the community, as they share information about how goats can help a family have a competitive advantage and generate additional income for the family.

Mrs. Khim Sonem's daughter is delighted with her pregnant doe and hoping to have more goats soon. ◆

THE LEARNING GUIDE

INTRODUCTION TO GOAT CARE AND MANAGEMENT

LEARNING OBJECTIVES

By the end of the session, participants will:

- Be acquainted with each other
- Be able to articulate their personal goal for the training
- Identify differences between milk and meat goats
- Demonstrate how to determine a goat's age by its teeth
- Demonstrate how to determine a goat's weight
- Be able to palpate a goat to determine Body Condition Scoring
- Understand how the ruminant stomach works

TERMS TO KNOW

- Rumen
- Palpate
- Chevon, Cabrito
- Body Condition Scoring
- Cud

MATERIALS

- Name tags – If a large button maker is available have templates for the name tag and then make a button for each person. Otherwise use heavy paper.
- Colored pencils or crayons
- A flip chart or a black board
- Three ropes or belts for holding the "goats" in skit
- Dairy and meat goats for the first activity
- At least three goats of different ages for the second activity
- A measuring tape for each participant
- A ruminant stomach from a slaughter house (if available)
- Handout of Body Condition Scoring Chart

ADVANCE ASSIGNMENT

- Select two class members to discuss the skit and present it to the group. Give them a copy of the sample script to read.
- Have two class members read the text on The Ruminant Digestive System and be ready to present this to the class using both technical and local names. They can use drawings or a ruminant stomach from the slaughterhouse.
- Prepare awards for participants who correctly identify the goat's age (optional).

TIME (May vary according to group)	ACTIVITIES
30 minutes	**Introduction** Welcome to all. Introduce the facilitators who will have already prepared a name tag. Each person will design his/her own name tag with a drawing of a goat on it. Get acquainted with one another by telling about the name tag, something about family and farm. (Two minutes each).
20 minutes	**Get Everyone Thinking and Talking About the Skit** See sample script at end of this learning guide of a theater skit about getting started in raising goats. In the skit people may be goats or they may be imaginary. Actors *ad lib* some of the conversation once they know the skit. Add or change conversation to discuss the challenges of raising goats in the community.
20 minutes	**Share Understandings From the Skit** ■ What happened here? What were Carmen and Maria talking about? What do you think of Carmen's plan – is she prepared to raise goats? What advice would you give Carmen before she starts raising goats? What steps should she take before she buys her herd? (This may lead to some of the activities below.) ■ What about this class? Who are we and why are we here? What are our expectations? ■ Why do you want to raise goats? (List the benefits and challenges.) How is goat raising different than what Carmen thought? How has it changed from how people raised goats in the past?
20 minutes	**Differences Between Dairy and Meat Goats** Have a dairy and meat goat at the front of the class. Ask several individuals to share differences.
20 minutes	**How to Tell a Goat's Age** ■ Talk about what happens when you buy a goat and don't know its age. ■ Practice identifying age of goats by their teeth. Participants approach goats in teams of two or three persons and identify age. There can be small rewards for those who are correct.
30 minutes	**Understanding the Ruminant Stomach** Two class members will present information on the ruminant stomach. Discussion questions: How does a goat's stomach differ from a human stomach? How does that affect the food eaten by each species?

30 minutes	**Determining the Body Weight of Goats** ■ Ask everyone to write down what they think each goat weighs. Ask the local way of measuring and weighing a goat. ■ Review and demonstrate the formula • identify heart girth • divide into groups, determine weight of goats and compare answers ■ Reflect: What are the benefits to weighing a goat this way? When will you use this information?
60 minutes	**Body Condition Scoring (BCS)** ■ Review BCS chart ■ Divide into teams of two with another team observing ■ Examine and palpate goat and note score ■ Switch teams and repeat ■ Bring teams together ■ Compare answers ■ Discuss: How is knowing the body condition useful to you? What is the best score? When will you use this method and information?

REVIEW-20 MINUTES

- What was useful?
- What was a surprise?
- What did not work out well?
- What do you know now that you did not know before?
- What can you do as soon as you get home?
- What practices will be difficult to do at home?

SAMPLE SCRIPT FOR GOAT RAISING SKIT

Chapter 1—Introduction to Goat Care and Management

Feel free to modify according to your context (change names, etc.)

Carmen meets Maria coming down the street with three goats.

Maria: "Look at those three lumpy old goats, Carmen. I didn't know you raised goats! Where are you going, to the vet?"

Carmen looks irritated: "No way, I just got a great deal at the livestock market - these are the start of my new herd. I decided yesterday that I want to raise dairy goats so I can send my daughter to school in the fall."

Maria: "So you think you got a good deal! Look at how that old buck is limping and look at the lumps on the front of that old doe's shoulders."

Carmen: "They aren't old! The lady who sold them said this one is three years old, and is a great breeder."

Maria: Looks at the goat's teeth and says – "She deceived you – these two goats are at least seven years old, and he's missing teeth and is as thin as a rail."

Carmen: "How can you tell? Are you a dentist? They are three quarter pure Landrace breed."

Maria: "Landrace? Anyway, Carmen, didn't you say you wanted to raise dairy goats?" Maria feels the muscles over the loin of one of the goats. "Hmmm, score of 2. Why did you get these skinny, short meat goats instead of some nice tall, Nubians with large udders from Sonia? And Carmen, what are you going to feed them?"

Carmen: "Sonia's goats eat too much feed. I plan to ask Mr. Julio if I can pasture them in that green grass next to the swamp across the street. No one uses that. Pablo can bring them over every morning before school."

Maria: "Pablo? He's only seven years old! And that's not grass, its just reeds, and full of snails. They're going to get liver flukes! Besides, there is no fence and if they get into Julio's vegetable garden, he'll be making goat stew. By the way, I never see you at the goat class that the extension service is sponsoring."

Carmen: "I plan to take the course next fall. We've raised hogs already, you know."

Marie: "Better teach those old goats how to go to the grocery store to buy food because there will be no milk from that doe."

The end. Goats and ladies take a bow.

THE LESSON

INTRODUCTION TO GOAT CARE AND MANAGEMENT

THE BENEFITS OF OWNING A GOAT

Human Nutrition

Goat meat, milk and by-products provide important nutritional components of a human diet. Goat meat, for example, has proven to be a low-fat, iron-rich source of protein. The United States Department of Agriculture has reported that the saturated fat in cooked goat meat is 40 percent less than that of skinless chicken and is low in cholesterol. Goat meat is 50 to 65 percent lower in fat than similarly prepared beef, but has comparable protein content. Protein in our diet builds strong bones, teeth, hair, muscles and antibodies to fight off infection. It is also required for the growth and repair of human tissue. Many people in developing countries have protein deficient diets.

Known as chevon or cabrito, goat accounts for more than 60 percent of the red meat consumed worldwide. Some 50 percent of the meat consumed by people in developing countries is goat. As little as 50 grams of goat meat per day will provide:

- All the protein needed for children from age 1 to 10
- All the protein needed for teens and adults
- Important vitamins and minerals
- Some 50 percent more iron than chicken

Goat milk and dairy products are also a major source of protein, as well as calcium in the human diet. Due to its smaller fat globules, goat milk is easier to digest than cow milk and is good for babies, children and adults. Calcium is an important mineral that strengthens bones and teeth. One liter of goat milk per day can provide:

- All the protein a child needs to age 6
- 60 percent of the protein a child needs to age 14
- Half the protein a child age 14 to 20 needs
- All the calcium a child to age 10 needs
- Almost all the calcium a child age 10 to 18 needs
- All the calcium elderly people need

Goat milk is lower in folic acid than cow or human milk. Therefore when fed to infants, goat's milk should always be supplemented with folic acid. Folic acid works with B vitamins, which help the body utilize protein and prevents anemia, poor growth and birth defects; it also supports the immune system.

Goat cheese and whey (a by-product of the cheese-making process) are rich protein and calcium sources. Whey also contains magnesium, phosphorus and potassium. It can be used in cooking to add vitamins and minerals to other foods. Cheese is an effective way to preserve and use excess milk.

Economic Gain

If given proper care, a goat can provide income from the sale of milk or meat. Below are examples of products that can be sold or used for food.

- One to four liters of milk every day
- Cheese and other milk products
- Kids to sell
- Kids to increase the number of goats in your herd
- Protein and iron-rich meat easily sold at the market or to neighbors
- Goat skins

Environmental Improvement

Manure and urine collected from goats can be used to improve the soil and make a garden grow better. Both can be put directly on the inedible portions of plants in the garden and mixed into the soil. DO NOT put manure directly on edible portions of the plants as this could contaminate plants with parasites and disease. Manure can also be placed in a bucket and covered with water overnight, creating a "tea" that can be put in the soil around plants. The mixture can restore soil quality and boost yields, but should not be put on growing foliage.

Another valuable source of fertilizer is whey, which can also be used as feed for pigs, chickens or goats. Besides calcium, magnesium and phosphorus, whey contains potassium, sugars and protein. Like manure, whey is a complete fertilizer and can be applied to the same crops.

Goats can be used to browse and clear land of unwanted species of plants. They can be used in land reclamation and stream bank enhancement, depositing manure and urine in these areas which helps new plants to emerge. By scavenging food wastes left along roadsides goats help clean up the environment. Care must be taken to assure that goats do not destroy crops and gardens.

CHALLENGES OF OWNING A GOAT

Raising goats for milk and meat has its challenges, especially in providing adequate food, water, shelter and health care for each goat.

- Feed goats plenty of green chop (forage) at least twice a day or take them to browse in pasture areas or along roadsides
- Provide salt, minerals and vitamins
- Give goats 6 to 8 liters of clean water every day
- Milk dairy goats twice a day, morning and evening, for maximum milk production
- Check daily for signs of illness. Be prepared to contact an animal health worker and to give various kinds of treatment and supportive care when needed.
- Goats require a secure enclosure to protect against predators and prevent unwanted damages to crops or vegetable gardens
- When tied or tethered, goats must be moved several times each day
- Beginning at two months of age, male goats must be kept separate from females to prevent premature or unplanned breeding

Despite the challenges of raising goats, there are many benefits of improved management. The table below illustrates the differences between management related to a local goat and an improved goat.

LOCAL GOATS	IMPROVED GOATS
Initial cost is less	Larger Investment
Scavenges freely	Feed is brought to goat
Selects diet from browsing	Selects diet from browsing or diet determined by owner and may include concentrates and purchased feeds
Salt/mineral block offered in pasture	Salt/mineral block offered in pasture or salt/mineral block offered in zero grazing pen
Uses trees and rocks for shelter	Uses trees and rocks for shelter or has shelter and pen built by farmer
Produces only enough milk for kids	Has milk for kids and milk for the family
Limited products to sell	Milk and milk products; live goats to sell
Smaller body size	Consumes more forage and supplements
Can be destructive to crops and gardens	Needs management to enhance land
Resistant to adverse conditions and may be immune to specific diseases	Managed health programs and higher risk of disease mean more veterinary costs if goats not acclimated.

PARTS OF A GOAT'S BODY

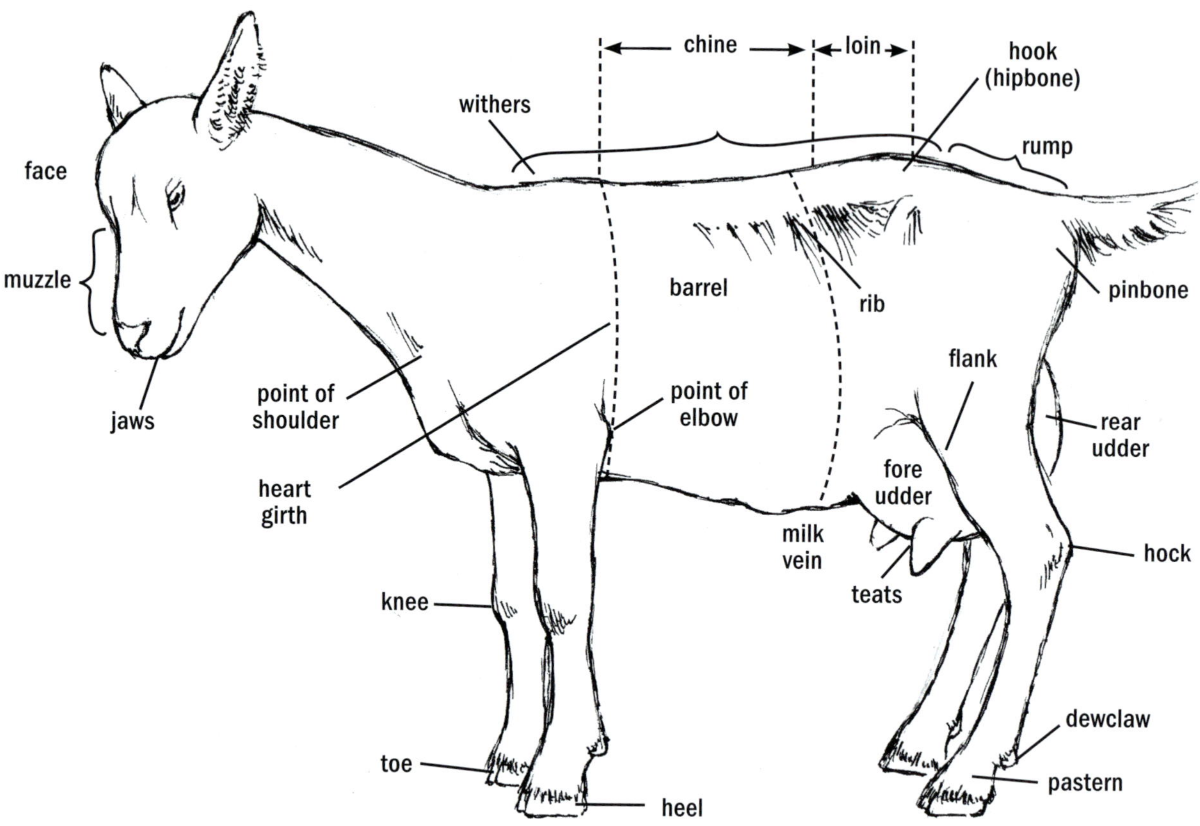

THE RUMINANT DIGESTIVE SYSTEM

Goats, like cows and sheep are ruminant animals, which have four compartment stomachs that allow them to digest roughage (food high in fiber) such as grass, hay and silage. In ruminant stomachs forages, corn stalks, banana leaves and agricultural by-products that are not digestible to single-stomach animals are transformed into nutrition for the goat. Single-stomach (monogastric) animals include humans, pigs and chickens.

The goat's stomach has four chambers: 1) the rumen, 2) the reticulum, 3) the omasum and 4) the abomasum. The size relationship of the four chambers changes as the animal matures. Goats grip grass or forage between the upper jaw palate and lower jaw teeth, ripping it with a jerk of the head. The grass is then swallowed whole and moves into the first compartment of the stomach – the rumen. The rumen, the largest compartment, contains many microorganisms which supply enzymes to break down fiber. It is often called the fermentation vat. The tiny organisms in the rumen help build proteins from the feed and manufacture all of the B vitamins needed by the goat. Vitamin K is also produced in the rumen.

When roughage is eaten by the adult goat, it is chewed, mixed with saliva and then swallowed. It goes down into the rumen where it is broken down or digested by the microorganisms. After the fiber is broken down, the grass or forage passes back through the reticulo-rumen and returns to the mouth to be chewed again for a longer time. This is called "chewing the cud." The entire process is called rumination.

The food particles then move to the second compartment, the reticulum, being separated only by a partial wall. Located just below the entrance of the esophagus into the stomach, the reticulum's lining looks like a honeycomb and serves to channel food and trap foreign material that might harm the animal.

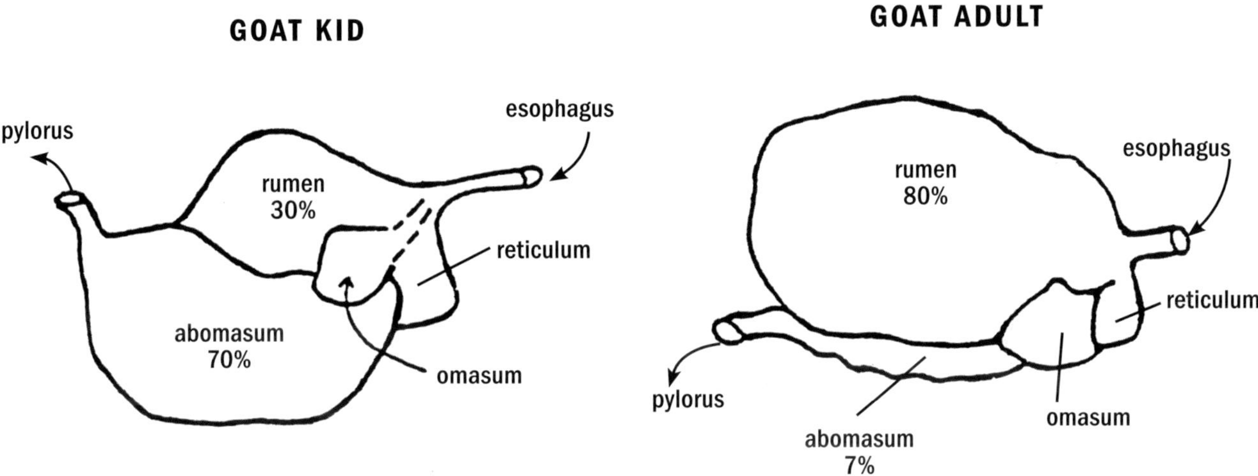

After passing to the reticulum, food moves to the omasum, which consists of hanging layers of tissue sometimes called the "manyplies." The large surface area of these folds permits absorption of moisture from feed and also absorbs more nutrients called volatile fatty acids that help supply the goat with energy.

The particles then pass into the fourth stomach, the abomasum. The fourth compartment is considered the true stomach. It contains hydrochloric acid and enzymes that break down feeds into simple compounds that can be absorbed by the stomach and intestinal walls.

For about six weeks after birth kids consume mostly milk that goes straight to the abomasum. As roughages gradually become part of the diet, the rumen enlarges to break down cellulose and the process of rumination begins.

DETERMINING THE AGE OF GOATS

When selecting a goat, it is helpful to know the age. This can prevent any misleading information in the purchase of a goat in the marketplace. A goat's teeth can be used to determine age. Goats have no front teeth in the upper jaw and eight front teeth in the lower jaw. Toward the back of the mouth goats have large teeth called molars to chew the grass.

Kids under one year of age have small, sharp front teeth. At age one, the center pair of teeth will drop out and are replaced by two large permanent teeth. At about two years, two more large front teeth appear, one on each side of the first two yearling teeth. The three to four year old has six permanent teeth. The four to five year old has a complete set of eight permanent teeth in front.

Be sure to check the teeth of an animal before purchasing. For five years and older, the approximate age can be determined by the deterioration of the front teeth. As the animal gets older, the teeth spread apart and finally become loose and eventually some drop out. At this age the animal begins to lose its usefulness as a grazing animal. It may be kept and fed specially prepared feeds if the animal is still capable of reproduction. When an animal looses its usefulness, it can be culled and sold for meat. The income can be used for younger replacements or for yearling goats.

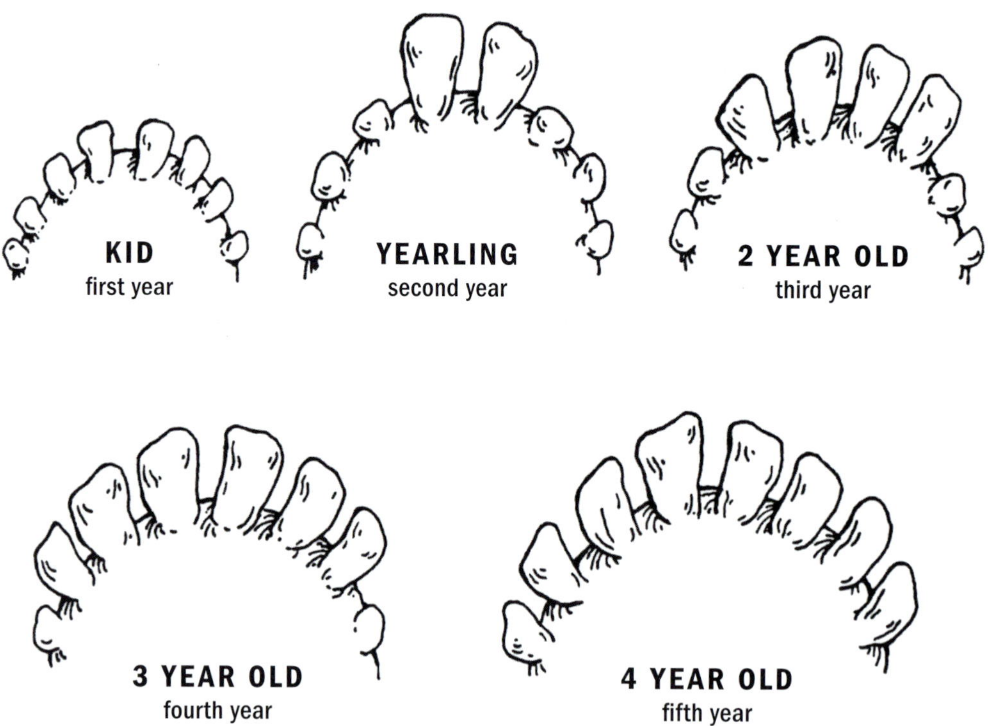

▲ Colby, Byron. *Dairy Goats Breeding/Feeding/Management*

ESTIMATING THE BODY WEIGHT OF GOATS

To determine the weight of a goat, measure the goat around the heart girth. Pull the tape tightly.

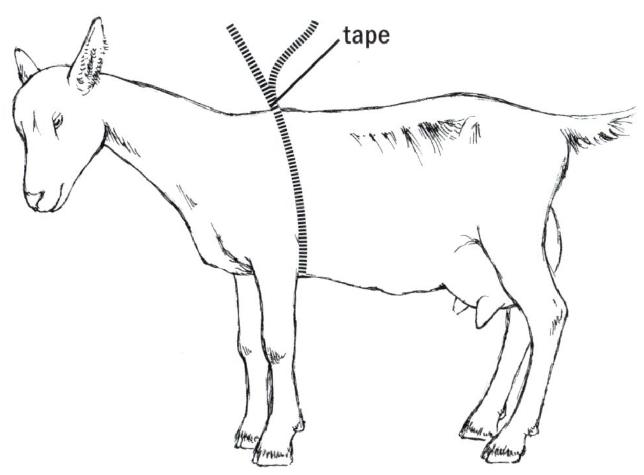

CENTIMETERS (cm)		INCHES (")	→	KILOGRAMS (kg)		POUNDS (lb)
27	=	10 ¾	→	2.3	=	5
29	=	11 ¼	→	2.5	=	5 ½
30	=	11 ¾	→	2.7	=	6
31	=	12 ¼	→	3.0	=	6 ½
32	=	12 ¾	→	3.2	=	7
34	=	13 ¼	→	3.6	=	8
35	=	13 ¾	→	4.1	=	9
36	=	14 ¼	→	4.5	=	10
38	=	14 ¾	→	5.0	=	11
39	=	15 ¼	→	5.4	=	12
40	=	15 ¾	→	5.9	=	13
41	=	16 ¼	→	6.8	=	15
43	=	16 ¾	→	7.7	=	17
44	=	17 ¼	→	8.6	=	19
45	=	17 ¾	→	9.5	=	21
46	=	18 ¼	→	10.4	=	23
48	=	18 ¾	→	11.3	=	25
49	=	19 ¼	→	12.2	=	27
50	=	19 ¾	→	13.2	=	29
51	=	20 ¼	→	14.1	=	31
53	=	20 ¾	→	15.0	=	33
54	=	21 ¼	→	15.9	=	35
55	=	21 ¾	→	16.8	=	37

CENTIMETERS (cm)		INCHES (“)	→	KILOGRAMS (kg)		POUNDS (lb)
57	=	22 ¼	→	17.7	=	39
58	=	22 ¾	→	19.1	=	42
59	=	23 ¼	→	20.4	=	45
60	=	23 ¾	→	21.8	=	48
62	=	24 ¼	→	23.1	=	51
63	=	24 ¾	→	24.5	=	54
64	=	25 ¼	→	25.9	=	57
66	=	25 ¾	→	27.2	=	60
67	=	26 ¼	→	28.6	=	63
68	=	26 ¾	→	29.9	=	66
69	=	27 ¼	→	31.3	=	69
71	=	27 ¾	→	32.7	=	72
72	=	28 ¼	→	34.0	=	75
73	=	28 ¾	→	35.4	=	78
74	=	29 ¼	→	36.7	=	81
76	=	29 ¾	→	38.1	=	84
77	=	30 ¼	→	39.5	=	87
78	=	30 ¾	→	40.8	=	90
79	=	31 ¼	→	42.2	=	93
81	=	31 ¾	→	44.0	=	97
82	=	32 ¼	→	45.8	=	101
83	=	32 ¾	→	47.6	=	105
85	=	33 ¼	→	49.9	=	110
86	=	33 ¾	→	52.2	=	115
87	=	34 ¼	→	54.4	=	120
88	=	34 ¾	→	56.7	=	125
90	=	35 ¼	→	59.0	=	130
91	=	35 ¾	→	61.2	=	135
92	=	36 ¼	→	63.5	=	140
93	=	36 ¾	→	65.8	=	145
95	=	37 ¼	→	68.1	=	150
96	=	37 ¾	→	70.3	=	155
97	=	38 ¼	→	72.6	=	160
98	=	38 ¾	→	74.8	=	165
100	=	39 ¼	→	77.1	=	170
101	=	39 ¾	→	79.4	=	175
102	=	40 ¼	→	81.6	=	180
104	=	40 ¾	→	83.9	=	185
105	=	41 ¼	→	86.2	=	190
106	=	41 ¾	→	88.4	=	195

DESIRED CHARACTERISTICS

Dairy Goat – Doe

A dairy doe should have:

- A bright, sleek, loose and pliable coat
- Alert, bright eyes with pink mucosa
- No discharge from eyes and nose
- Feminine characteristics, including fine features
- A straight top line and sharp withers. One should be able to feel individual vertebrae along the spine.
- A long rump that is not too steep and not too straight
- A wide chest
- Plenty of capacity for feed and kids; spring of ribs
- A deep heart girth
- Straight legs with prominent hips and sharp pins.
- A strong muzzle with a jaw that is not undershot nor overshot
- Good teeth that are not loose, broken or missing
- A sound walk with no trace of a limp
- An udder that blends smoothly with the body wall and is securely bound across the entire upper surface with two teats. The udder should be carried well forward, sloping gently into a gradual upward curve. The rear udder attachment should be high, wide, firm, strong, and firmly fastened to a wide escutcheon.

Here are two examples of healthy udders:

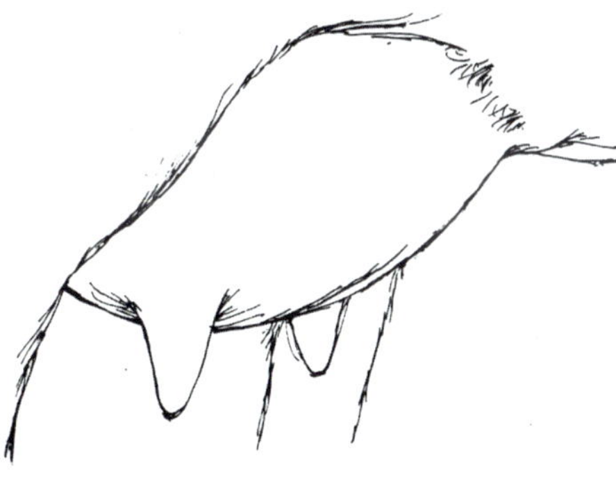

Here are examples of poor or unhealthy udders:

Dairy Goat – Buck

The buck used for breeding is one-half your herd. Therefore, it is important to look for a buck that will reinforce the doe's strong genetic points and does not share any of her major genetic flaws. Below are a few indications of a suitable breeding buck:

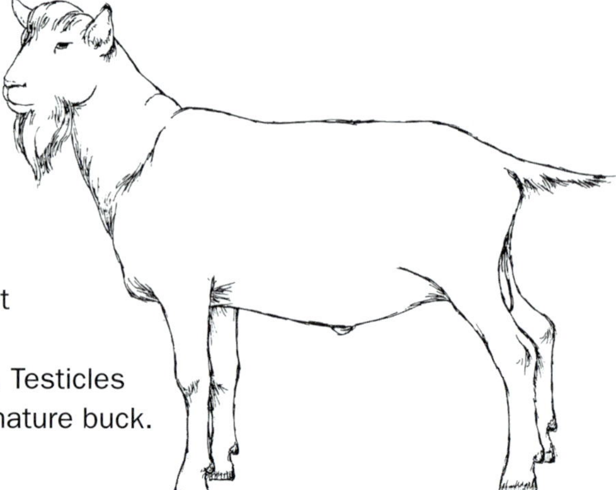

- Excellent health and sexually virile
- A masculine body with medium-length head
- A strong, broad muzzle with large open nostrils
- Bright eyes with pink mucosa
- No underbite or overbite
- A strong, straight, smooth back
- A long, wide and nearly level rump
- Strong, sturdy legs that are wide apart and squarely set
- Solid feet
- A pear-shaped scrotum with two testicles of equal size. Testicles should have a circumference of at least 29 cm in the mature buck.
- A deep heart girth and wide chest floor

Meat Goats

In addition to similar characteristics of dairy does and bucks, a quality meat goat should have:

- Structural correctness, but with an emphasis on muscle volume. Muscling visible in forearm, hindquarter and inside rear legs. Muscling evident in a thick thigh and the depth of the twist. (The twist is the area between the bottom of the anus and where the legs split. Goats that are deep in the twist have more leg muscling.)
- A strong, wide and level back with a sturdy wide rump and loin. The width and length of the loin are correlated to the volume of meat in the carcass.
- The rump should be long with a slight slope from hook bones to pin bones, but not overly steep. A five to seven degree angle is ideal.
- Meat goats should be above average in overall length of body and general size
- The front end, like dairy does and bucks, should be wide and smooth with well spaced front legs and a broad, deep chest floor.
- The cannon bone should have an adequate length from knee to pastern. A long body is desirable with bone length and cannon bone diameter in proportion.
- Strong legs and sturdy feet are essential. Rear legs should be straight and set wide apart. Pasterns will ideally be strong.
- Widespread toes for ease in climbing and walking on hillsides and rock outcroppings

Meat Goat – Doe

- Smooth shoulders that blend smoothly into the neck and rib cage (as opposed to pointed, like dairy does)
- Flat over withers
- The body should have volume and capacity, demonstrating productivity to breed, carry and rear young in an extensive pasture situation. The body should be of adequate size for breed and age.

- Does should exhibit good spring of rib and depth of body; these are good indicators of volume
- The udder should ideally be round, symmetrical and situated well above the hocks. The udder should have good suspension (not pendulous) and two teats that are easily nursed by a newborn kid.
- There should be adequate muscling in the rear legs without loosing femininity

Meat Goat – Buck

- Meat bucks should exhibit masculinity and adequate muscling
- The head should have a broad strong muzzle and horns set far apart
- The body should demonstrate the masculine profile with a heavier chest and fore body, and often coarse shoulders as a manifestation of testosterone.
- Testicles should be of equal size, smooth, lump-free, and between 29 cm to 32 cm in circumference
- Buck should have only two rudimentary teats

BODY CONDITION SCORING

Body Condition Scoring (BCS) is an important monitoring tool. Scoring provides a starting point for assessing and managing goats to improve performance, limit health problems associated with improper nutrition and increase profits. BCS is a measure of energy reserves and allows one to estimate changes in energy balance that lead to weight gain or loss. This is especially important in breeding programs. A higher BCS at kidding and early lactation will improve the success of the next kidding season by increasing the rate of ovulation and conception. Furthermore, adequate energy stores at parturition ensures higher milk production for a longer period of time. Thin (low BCS) does often are late kidders with greater likelihood of abortions, fewer or weaker kids per birth, and lower milk production.

Body weight is made up of several things including: breed, body conformation, frame size and mature size, stage of gestation and stage of lactation. As an adult, the body frame will remain constant, but weight will change based on deposition of fat and muscle. The amount of fat and muscle depends on the nutritional and physiological status of the goat. Do not include the fetus or associated fluids when evaluating weight changes. Be careful when culling on BCS at weaning as looks can be deceiving. It is especially important to palpate the animal to assess the BCS. An animal may look like it has adequate fat reserves because of the fetus, but may still be underweight. Use the weight tape to measure the body weight.
The most important times to make a BCS assessment are the last three weeks of gestation, six weeks into lactation, at weaning (cautiously) and before breeding. Be sure to physically palpate the goat to get a BCS. Simply looking at the animal is not enough. The best places to feel are the rib and along the sides of the spine.

AN PEISCHEL

▲ *Palpating a Goat.*

It is best for goats to have a BCS between four to six (moderate). These goats will consistently perform to optimum production. A drop in body condition should be gradual and regained rapidly. This group should be flushed at five to attain a six at breeding and kid at five or six. They should not drop below a four by the end of lactation which can be maintained by providing adequate nutrition. It takes four to six weeks with a supplement (or changing to higher quality forage) to return a goat with a BCS of three to a four.

Thin goats (score of one to three) have: poor reproductive success, lower twinning rates, decreased weaning rates, increased chance of internal parasite load and lethargy. Severely restricted nutrition of the doe will greatly depress the weaning weight of the kids. Therefore, under poor management and adverse environmental conditions, kid performance can be enhanced by managing the nutritional level of the kids separately. When BCS decreases, it is time to increase the amount of feed or to move the animal to an area with higher quality forage.

Fat goats (score of seven through nine) will be more prone to pregnancy toxemia, fatty liver syndrome, kidding difficulties, lower conception rate, impaired immune response, less energetic browsing/brushing and reluctance to travel long distances over rough terrain. Fat goats indicate you are wasting feed and money.

SUMMARY–BODY CONDITION SCORING

Body Condition Scoring Chart

Objectives

- To monitor the nutrition program
- To minimize internal parasite problem (in addition to fecal checks)

Areas to be monitored

- Tail head
- Ribs
- Pins
- Hooks
- Edge of loin
- Shoulder
- Backbone – transverse and spinous processes of vertebrae
- Longissimus dorsi muscle
- Sternum

Scale

- Thin: 1 to 3
- Moderate: 4 to 6
- Fat: 7 to 9

Appropriate BCS

- End of pregnancy: 5 to 6
- Start of breeding season: 5 to 6

Recommendations

- Animals should never have a body condition score below 4
- Pregnant does should not have a BCS of 7 or above at end of pregnancy due to risk of pregnancy toxemia. A BCS of 5 to 6 at kidding should not drop off too quickly
- Maintain a moderate BCS at all times: 4 to 6

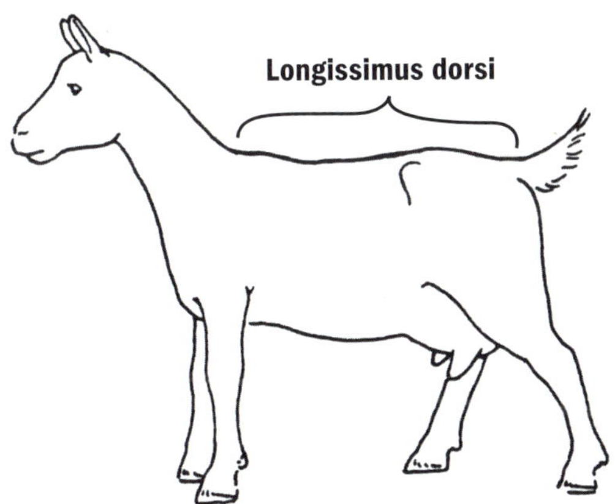

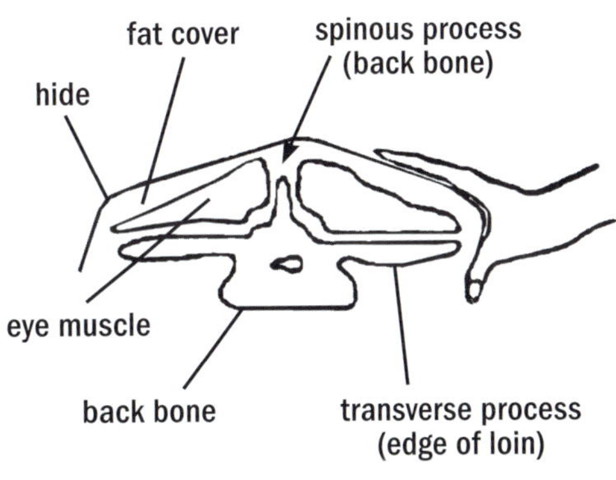

BCS 1	■ Extremely thin ■ Emaciated	Extremely thin and weak, near death. Outline of ribs visible and spinal processes distinct and prominent with severe depressions, physically weak; shoulder, loin and hindquarters atrophied in appearance, skin adheres to bone.
BCS 2	■ Extremely thin	Extremely thin but not as weak or emaciated as BCS 1. Skin in direct contact with bone; prominent "V" shaped cavity under tail, outline of spine and ribs visible, bony surface of the sternum protruding.
BCS 3	■ Very thin ■ Frame visible	Wasting in appearance. All ribs visible. Spinous processes prominent and depressions obvious (rib, hips); sunken between pins and hooks; sternum is prominent. No fat cover felt with some muscle wasting.
BCS 4	■ Moderate ■ Slightly thin	Some ribs visible. Spinous processes and sharp. Individual processes can be easily felt. Thin flesh covering hooks and pins. Definite depression between hooks.
BCS 5	■ Moderate ■ Frame covered ■ Balanced	Spinous processes felt but are smooth; transverse processes have smooth concave curve; hooks and pins smooth; muscle becoming obvious, Sternum can be palpated.
BCS 6	■ Good ■ Slightly fleshy ■ Smooth cover	Smooth look with ribs not very visible. Spinous processes smooth and round. Spinous to transverse processes smooth sloped. Individual processes very smooth, felt with considerable pressure. Slight depression between hooks and pins.
BCS 7	■ Fat ■ Frame not visible ■ Fleshy	No spinous processes noticeable; ribs not visible, spinous process felt under firm pressure. Hooks and pins rounded with some cover; flat between hooks; palpation of sternum difficult.
BCS 8	■ Fat ■ Obese	Animal is very fat with spinous processes difficult to feel. Ribs can not be felt. Animal has blocky obese appearance; tail-head cavity filling with fat.
BCS 9	■ Fat ■ Extremely obese	Severely over-conditioned. Spinous processes buried in fat. Similar to an eight but more exaggerated. Animal has deep patchy fat over entire body. Tail-head cavity exhibits fat filled folds.

J.M. LUGINBUHL, M.H. POORE, J. P. MUELLER, AN PEISCHEL

HOOF TRIMMING

While it is possible for goats on a rough terrain to wear down their hooves, most will need trimming at regular intervals. The diagrams below show the proper trimming method. Use either the Burdizzo Hoof Trimming Shear, a garden pruning shear or a sharp knife. If hooves are trimmed too close by mistake and blood appears, use an antiseptic.

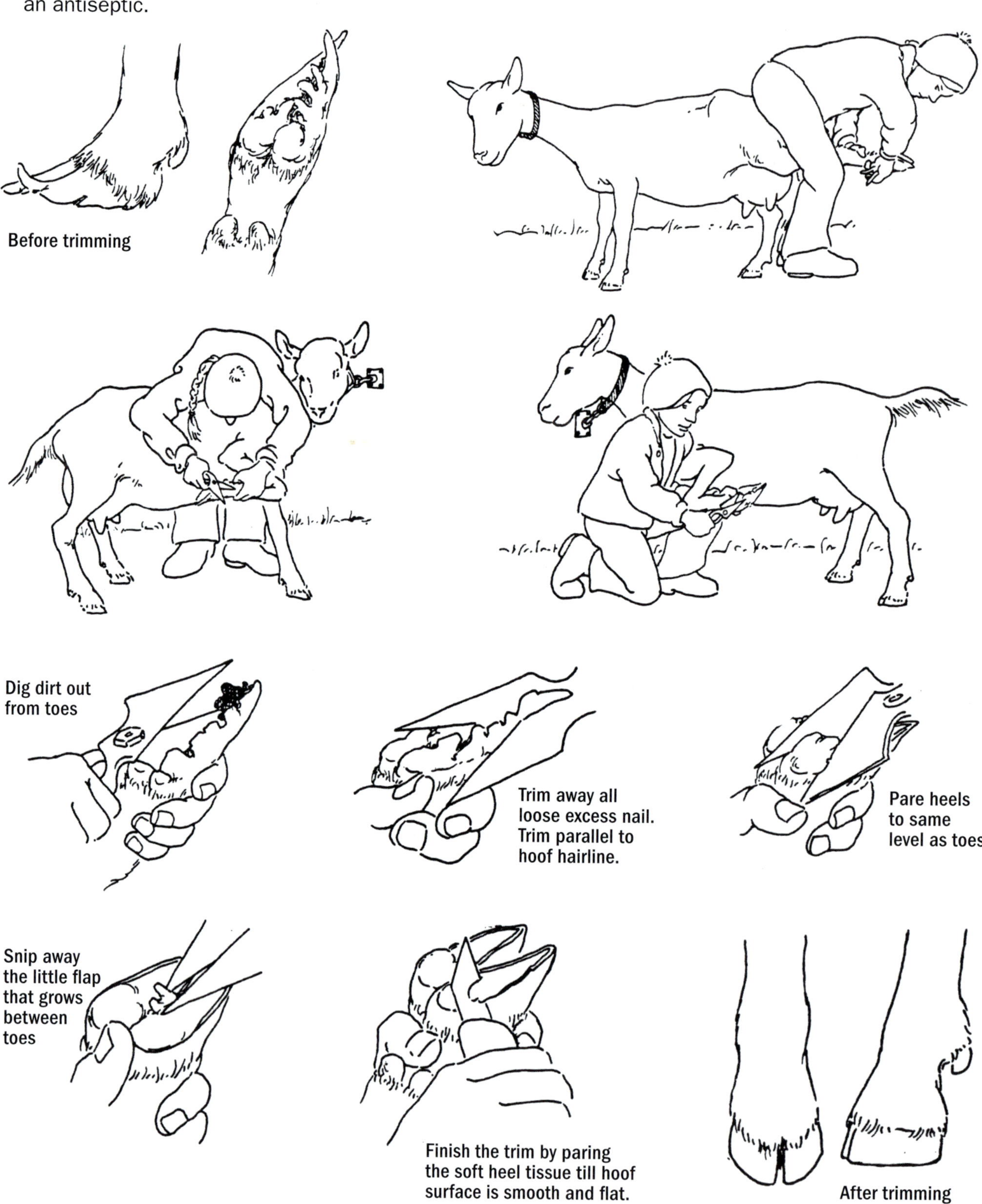

IDENTIFICATION OF GOATS

Identifying goats is central to effective management. Many farmers do so by simply naming their goats, which often respond when called. While recognizing each individual goat is easy in a small herd, permanent identification will be needed as herd numbers grow.

Most goatkeepers use one of two methods of identification: tags or tattoos. Tags are placed either around the animal's neck or in the animal's ear. Tattoos are normally placed in the ear or on the tail web. Each method has its advantages and disadvantages. Use the table below to select a method of identification.

ROSALEE SINN

▲ *Ear Tag*

ROSALEE SINN

▲ *Neck Tag*

IDENTIFICATION METHOD	LOCATION	ADVANTAGES	DISADVANTAGES
Neck tag	On cord or chain collar around neck	■ Requires no specialized equipment other than the tag itself. ■ Break-away collars provide for safety from entanglement.	■ Break-away collars may result in unidentified goats and lost tags. ■ If a goat should become entangled, chain or sturdy cord collars could result in strangulation.
Ear tag	In ear	■ Easily applied. ■ Large tags are easily visible once in place. ■ If governmental requirements specify ear tags for disease control purposes, these can double as ID tags.	■ Requires specialized applicator. ■ Goats are especially prone to ear tag tear-out. This results in lost identification and damaged ears. ■ The weight of a large tag may distort erect ears or cause a wound that can lead to flies or maggots.
Tattoo	In ear or tail web	■ Not susceptible to tear out. Most permanent of the three ID methods.	■ Requires specialized equipment. ■ Tattoo can be read only on close inspection. Ink can fade over time and environmental extremes can make a tattoo difficult to read.

Empowering Women through Small-scale Goat Farming

A village housewife in Cameroon's Northwest Province, Nkume Margaret had little education. Her husband strictly controlled her activities. She spent much of her time on strenuous farm work, and housekeeping, caring for her husband and four children. Her life was very difficult. There was not enough food for the family and although the children attended school from time to time each wave of new school fees drove them back home.

Nkume Margaret attempted smallholder pig farming, but because she lacked appropriate training her work yielded few dividends. When she learned about the goat project she saw a solution to many of her problems. She contacted Heifer Cameroon through the Chuketam Hunger Fighters Farming Group and asked for assistance. She took the training course. She learned about using goat manure to enrich her garden and how to market her extra goats. Through all of these activities she could earn money for school fees and supplies. And goat meat would provide much needed protein for her family. A successful venture would provide her with a measure of independence she badly needed.

Nkume Margaret's husband and children assisted in the preparation of a shelter and planting forage. She received five goats from Heifer Cameroon. Her children and husband help with the goats and she has become very successful. She has sold three goats for about US$54 each and has built her herd to 11 goats. With proceeds from that first sale, she sent her son to study engine mechanics in the provincial capital of Bamenda.

With money from the sale of other goats and money earned from vegetable sales from her now-thriving garden, she spent about US$500 to buy a grain mill. Within two months of buying the mill, she made a gross profit of $210. She used much of the remaining money to sponsor her children in school. In addition she deposits her weekly profit of US$36 in the local credit union.

▲ *Nkume Margaret with her goats*

Nkume Margaret sold last season's harvest for US$300 and used the money to buy three loads of sand for use in building a new house.

With training, a steady entrepreneurial hand, smart investments and hard work, since 2005 Nkume Margaret has increased the family's level of food production. She no longer needs to buy basic foods from the local market because she produces them herself. Thanks to the extra income, her children are doing well in school and her life has changed for the better. She shares the knowledge gained with her husband and children on issues like gender, participation, simple accounting methods, cleanliness and the prevention of HIV/AIDS and other diseases.

She advises her fellow villagers to be part of the Heifer International farmer's team because the assistance will always be useful at any moment of their rural life. Nkume Margaret looks into the future with the hope of becoming self-sustainable. ◆

THE LEARNING GUIDE

HUSBANDRY SYSTEMS, SHELTERS, FEEDERS AND EQUIPMENT

LEARNING OBJECTIVES

By the end of the session, participants will be able to:

- Describe the advantages of different goat management systems
- Determine the system that best fits their community
- Identify ways to improve their current management practices
- List ways in which goat farmers can work together for success

TERMS TO KNOW

- Husbandry
- Extensive production
- Semi-intensive production
- Intensive production
- Zero grazing
- Household goats

MATERIALS

- Paper and Pencils
- Large sheets of newsprint or paper
- Drawings for *Find the Safe Goat* Activity
- Cameras
- Photos of farms in area

ADVANCE ASSIGNMENT

- Select three class members to take photos of their farm and be prepared to describe their management practices to the class.
- Select two farms for the group to visit.
- Ask four participants to prepare drawings for *Appropriate Goat Farm* Activity. (Alternatively, they can do it during the session as part of the activity.)

TIME (May vary according to group)	ACTIVITIES
20 minutes	**Group Sharing** Selected participants will show photos of their farms and describe.
90 minutes	**Get Everyone Thinking and Talking** Divide group. Make visits to several farms. Ask participants to list the advantages and disadvantages of what they observe. Report back to group.
30 minutes	**Find the Safe Goat** ■ Participants divide into small groups and are given one of the drawings to discuss: • Goat tethered, tight rope wrapped around pole. • Goat grazing at night, predator watching. • Goats huddled in the rain, no shelter. • Goat in zero grazing pen with forage. ■ Report back to group regarding the illustration assigned; why it is or is not safe and how the situation could be improved ■ In the large group, facilitate a discussion about: • different husbandry systems and their advantages. • hazards of local predators, tethering and wet weather.
30 minutes	**Design Appropriate Goat Farm for Area** ■ In small groups, have participants draw a picture of a typical goat farm. The drawing should include typical shelters, equipment, fencing, grazing areas, etc. ■ Discuss the various drawings and the approximate cost to implement each of the designs.
15 minutes	**Brainstorm Ways in Which Farmers in the Community Can Work Together**

REVIEW-20 MINUTES

- What was useful?
- What was a surprise?
- What did not work out well?
- What do you know now that you did not know before?
- What can you do as soon as you get home?
- What practices will be difficult to do at home?

THE LESSON

INTRODUCTION TO GOAT CARE AND MANAGEMENT

The first consideration in goat raising is determining which system of management is best suited to the farmer's objectives and herd size. These decisions will also determine the type of goat that will be the best suited for the particular situation.

HOUSEHOLD PRODUCTION

Goat raising is an important component of household livelihoods for small farmers in developing areas of the world. Goats can be integrated into small scale farming systems for improved use of inputs and more efficient recycling of farm resources. One or two goats and their offspring can provide milk and meat for the family and offspring and surplus products to sell for income. Since goats breed and produce in a comparatively short period of time, income for basic household needs, school fees and better health care can become available quickly.

HUSBANDRY SYSTEMS

The main goat husbandry systems are: zero grazing, semi-intensive and extensive.

Zero Grazing

Often used in areas where land is in short supply, zero grazing is the most labor-intensive form of goat husbandry. A zero grazing unit is a covered pen, often with raised slatted floors, where animals are kept most of the time. Where small numbers are concerned, it is a more economical use of labor than herding the goats in the bush.

In this system, fodder must be cut and carried to the animals along with fresh water twice a day. Manure must also be removed daily. It is important that the zero grazing unit have an exercise yard and access to sunlight.

The advantages of zero grazing are:

- Keeps goats from destroying gardens and other crops. Reduces the chance of erosion from overgrazing.

TERRY WOLLEN

▲ *Indonesia – zero grazing pen.*

- Utilizes crop residues, vegetable peelings and other agriculture by-products for feed
- Is well-suited to high-producing animals, exotic breeds and their crosses which are more susceptible to local diseases
- Ideal for dairy goats under tropical conditions and perfect for small family herds
- Allows for easy collection of manure and urine for fertilizing crops
- Allows animals to develop a desirable bond with the farmer
- Allows for closer observation of animals, so heat (estrus) and mating can be controlled
- Allows farmers to notice and control disease earlier.
- Protects goats from predators
- Reduces tick and parasite numbers since the animals do not graze on infested pastures

Semi-Intensive Systems

This term covers goat raising practices between extensive management and zero grazing. Semi-intensive systems usually involve controlled grazing in fenced pastures, supplementary concentrate feeding and tethering. Animals are penned at night, usually nearer to home, to protect them from predators and bad weather.

The great advantage of permitting some grazing is that it gives the goat an opportunity to supplement its diet and to do some selective feeding to overcome dietary deficiencies. It is, however, important to avoid heavily used or contaminated areas. Be sure to have legal access or permission to use grazing areas.

In these systems, use care when tethering goats to prevent strangulation and protect against predators. Shade and clean drinking water must always be available. Shelter from rain and sun must be provided. It is essential to change the place of tethering every day so that fresh herbage and a variety of plants can be obtained by the animal.

Extensive Production

Extensive husbandry systems usually involve larger numbers of animals and a broader land base. These systems are rarely used with dairy goats, but are common for meat and hair goats in grassland regions of the world.

Extensive husbandry systems usually involve a human herder (goatherd), or at least a guard animal such as a dog, llama or a donkey. The herder is usually a young boy or girl whose task is to keep herds separate and out of cultivated land, as well as to identify sick animals and ward off predators. Grazing goats remove weeds and deposit manure, but can damage crops. To protect rangeland, farmer groups must work together to rotate pastures and avoid overuse of land. One of the considerations is how graziers (herders) and crop farmers work together. In regions where climatic conditions are favorable and predators are not too numerous, animals may be left free to graze on natural pastures without fencing or housing, being herded into enclosures only when required. The most extensive system in the tropics is bush or range grazing with housing at night and during rain showers. This system is often used when mixed-species i.e. goats, sheep and cattle graze together.

SHELTERS AND HOUSING

Goats, although hardy in other ways, are generally intolerant of wet or damp conditions. Under such circumstances they are stressed, making them particularly susceptible to pneumonia and other diseases.

The type of shelter, stilted or ground level, will depend on several factors: size of herd, cultural preferences, climate and choice of grazing/browsing systems. Housing should provide an environment that has good ventilation, is dry, free of drafts, and protects animals from sun and rain.

In all housing or shelters, each goat needs:

- 1.5 meters by 1 meter of space (mature goats)
- Protection from rain, heavy winds, cold and too much sun
- A dry floor (well-drained soil works well)
- A pallet or sleeping bench
- A place to exercise with rocks or wood constructed climbing areas
- Feed and water available on an as needed basis
- Protection from dogs, other predators and thieves

Stilted Housing

This type of shelter is common where rainfall is heavy and goats need protection against mud, standing water or flooding. In this type of housing the floor of the pen is raised about 1 to 1.5 meters above ground level and slatted floors are used. Slatted floors should be wide enough for manure to fall through, but narrow enough so that a goat's legs do not get caught. The height of the pen floor should facilitate removal of manure, Wood used for flooring should be flat or regular, not raw or unfinished, or goats will develop malformed feet.

Ground Level Housing

Ground level housing with an earthen floor made of packed clay or dirt takes several forms. One is the lean-to type of building pictured here. Drainage ditches may be installed around the ground-level housing to keep the floor dry during heavy rains. These ditches can lead directly to a compost pit. Pallets and sleeping benches will also keep the goats dry. A few rocks in the area provide a place to climb and to keep hooves trim.

In all shelters, manure and urine must be removed on a regular basis to keep the shelter clean.

MANGERS AND OTHER FEEDING EQUIPMENT

The most important function of mangers and other feeding equipment is keeping the green chop, hay or vegetable cuttings off the ground to prevent contamination from soil, feces or parasites. There are many different kinds of mangers, but all should be built according to the goat's size.

A good manger:

- Allows a caretaker to feed without entering the pen (This is sometimes difficult to achieve)
- Prevents goats from soiling the feed with dirt and manure
- Protects feed from rain
- Is appropriate to the age and size of the animal

Creep Feeder

In zero grazing shelters, it is important for kid goats to have access to small mangers and feeders without competition from older goats. Creep feeders consist of a small penned area with an opening only large enough for kids.

Slatted Feeder

Slatted feeders consist of a V-shaped rectangular feed box covered with wooden slats on each side of the V that serve as a receptacle for hay. The hay is available to the goats between the slats. Make the trough of the feed box at the bottom wide enough to hold any hay that may drop. Placing a bar across the top can prevent goats from sleeping in the feeder. As dust from the hay can cause eye irritation, slatted feeders should not be placed too high above a goat's head. If the slatted feeder is not inside of a shelter, a roof over the feeder may be needed.

Keyhole Feeder

Keyhole feeders, as the name implies, are structures attached to fencing that allow a goat to eat only when its head is inside the key-hole shaped opening. As forage is stored on the other side of the opening, feed cannot be contaminated. The openings on these feeders can be made to accommodate mature goats or kids. The caretaker can feed without going into the pen.

Hay or Forage Can Be Tied

In the absence of other feeders, hay can be tied to the pen or on a tree branch to keep the feed off the ground. Remember that dust can cause eye irritation, so feed should not be tied too high above the goat's head.

WARNING

Goats can be strangled by the rope after the hay or forage has been eaten. Tie the rope carefully to prevent the goat from getting caught in the rope.

Water Buckets

Water buckets should be clean and filled with fresh water twice daily. Place water buckets low enough to be reached by small animals, but high enough to prevent smaller animals from falling into the bucket and drowning. Hanging buckets keeps the water from being soiled. Water buckets can also be placed outside the pen with access through keyhole openings.

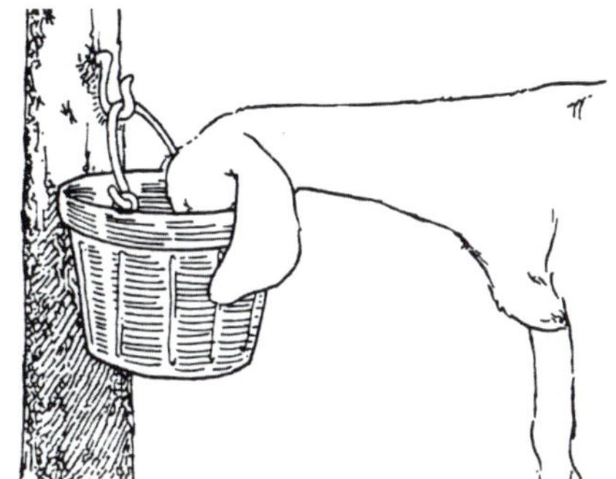

Salt Boxes

Although some farmers prefer a mineral block, loose mineral salt is best. Minerals should be offered freely and like feed be kept off the ground. The container can be a small wooden box or a dry hollow gourd suspended by a rope. If mineral salt is not available, animals can be given common salt with iodine. The mineral salt should be offered within a covered pen or be protected from rain.

FENCING, TETHERING AND HERDING

Fencing

Goat farmers utilize fencing to control the movements of the animals, separate does from bucks, protect croplands and block predators There are many kinds of fencing made from different types of material: living, sapling, wire, electric and solar. Choosing the right fence will depend on budget and purpose. Regardless of the material, fences must be designed to prevent goats from jumping or climbing over them and should guard against predators. Do not use barbed wire.

Living fences are planted by farmers to contain goats. *Gliricidia sepium* and *Erythrina berteroana* are examples of fodder trees that can be used for living fences with the added benefit of providing feed and firewood. Sapling fences are made from cut saplings laced together. Another option is wire fences, which are strong, but expensive. Goats can also be controlled with electric fencing, that can be solar-powered or battery-powered.

▲ *Electric Fence - Crystal Brook Farm - Sterling, Massachusetts*

▲ *Sapling Fence - Rwanda*

▲ *Living Fence - Rwanda*

Tethering

Ties or tethers are used to restrict the movements of a goat to a shaded or browsing area. A running tether consists of a long wire (or rope) that is staked into the ground at both ends and attached to the tether with a ring. As the tether moves across the wire, the goat can forage in a wider, but controlled area. Overhead tethers, in which the goat is tied to something above the animal's eye-level, prevent the goat's legs from tangling in the rope. The swivel tether is also valuable in preventing injuries as the tether rope swivels, rather than twisting around the animal.

▲ *Swivel Tether*

WARNING

Do not tether goats in an area where dogs or wild animals may attack. Goats should be watched carefully the first few days they are on a tether to make sure they do not panic and strangle while getting used to being tethered. Be sure to check for any poisonous plants in the surrounding area before tethering. Avoid tethering your goats where other goats and animals have grazed as they may pick up parasites. Make sure plenty of water is available for animals left unattended.

Herding

There may be times when you want to herd your goats so they can get different varieties of forage.

Success Through Leadership and Sharing

▲ Doña Antolina Serech López with her family and goats.

Doña Antolina Serech López, 48, and her three small children live in Chivarabal village in the Tecpán municipality in Guatemala. Her native language is Kaqchiquel, but she has learned some Spanish in order to communicate with institutions and get assistance for her group.

Doña Antolina is very active in her community. In November 2003, she organized a group of eight village women to participate in the FUDI Dairy Goat Project which is sponsored by Heifer International and the Integral Development Foundation (FUDI)

She has received two goats from the project. Since then, she has raised three offspring, a male and two females. She has participated in "passing on the gift" once and is ready to pass-on another offspring. She is waiting for a family close to her because she is so attached to her animals that she does not want them to go far away; she wants to see them grow.

Every day she provides her children with milk from her goats. She uses goat manure to enrich the soil in her squash garden. Also, with her husband's support she has been able to pack processed goat manure in sacks to sell in the community for US$4 a sack. She has planted trees like aliso (llamo) and palo de pito that grow quickly and provide excellent feed for her animals. When the trees mature she uses the wood for cooking fires.

Doña Antolina has a veterinary kit and has learned how to vaccinate the animals. Her group received the kit so everyone could benefit. The prices for the medicine are much lower when compared with the products sold at clinics. Doña Antolina set up a revolving fund with the money that every project participant pays to buy new medicine when needed. She has learned to treat common illness in goats and to give recommendations when an animal gets sick. She has also learned to use herbs and home remedies. When she doesn't know what to do, she consults with the technician in charge of the project. With his recommendations, she has been able to deal with infections, assist in births and care for newborns.

In addition to Doña Antolina's work on the goat project she also directs two other important group activities. One produces homemade jelly using local fruits, such as elderberry, apple and peach, to sell in area stores, supermarkets and coffee shops. The other project produces güipiles, handmade traditional blouses, for sale in the local market.

She shares her knowledge and experience with the other group members and facilitates training in her native language. Doña Antolina and the group are working on integrated production to diversify family incomes and make them more sustainable. Under Doña Antolina's leadership the membership has doubled. Her unselfish sharing and devotion to the community is an example for many of the Chivarabal women. ◆

THE LEARNING GUIDE

NUTRITION AND FEEDING

LEARNING OBJECTIVES

By the end of the session, participants will be able to:

- Describe the five major classes of nutrients required in a goat's diet
- Explain what is unique about the ruminant digestive system
- Plan for appropriate ways to care for and feed pregnant does, newborn kids, weaned kids, meat goats, bucks and lactating does
- Relate feeding practices and health needs to the body condition scores

TERMS TO KNOW

- Ruminant digestive system
- Total Digestible Nutrients (TDN)
- Protein
- Energy
- Forage
- Concentrate
- Fiber/Roughage
- Colostrum
- Immunoglobulins/Antibodies

MATERIALS

- Samples of local forages, cereals, concentrates, mineral salt
- Cards with the words colostrum, milk, water, protein, energy, fiber written on them. Make four cards of each word
- A large mural showing goats at different stages of development
- Farm Information Form at end of learning guide as handout

ADVANCE ASSIGNMENT

- Volunteers from the class will draw a mural of goats at different stages: kid nursing, buck, milking doe, pregnant doe and meat goat. Show goats in zero grazing pen, browsing, in pasture. Label each.
- Gather samples of local forages, cereals, concentrates and mineral salt.
- Two members of the class should prepare to share information about the ruminant stomach. If they can get a stomach from the slaughterhouse this can be very helpful.

TIME (May vary according to group)	ACTIVITIES
10 minutes	**Group Sharing** Greet participants and ask if they have anything to share or any questions.
20 minutes	**Get Everyone Thinking and Talking** Talk about the foods that are the staples in the community and whether or not people have a selection. Do you determine your food by availability, cost, taste or cultural preferences? Do you think about the nutritional value of the foods you eat? Do you eat mostly foods that are high in protein, energy, vitamins, minerals, fat or water? What happens if you miss any one of these for very long?
10 minutes	**A Goat's Diet** ■ How is a goat's stomach different from a human stomach? How are they the same? What five nutrients do goats need? What other things should be considered? ■ Describe the functions of the various compartments of the ruminant stomach.
15 minutes	**What is Grown on My Farm?** ■ Fill in the Farm Information Form.
15 minutes	**Work on a Mural Together Showing the Care and Feeding of:** ■ Pregnant does ■ Newborn kids ■ Weaned kids ■ Meat goats ■ Lactating does Have each individual place names and feed samples by the goat that needs this feed, liquid or mineral.

REVIEW-20 MINUTES

- What was useful?
- What was a surprise?
- What did not work out well?
- What do you know now that you did not know before?
- What can you do as soon as you get home?
- What practices will be difficult to do at home?

FARM INFORMATION FORM

- A pregnant goat needs a diet that is 12 percent protein and provides carbohydrates for energy. Consider the protein in the forage you are feeding and supplement with concentrate, minerals and plenty of fresh water.
- Kids, at one hour to three days of age, need: colostrum and milk from the dam.
- Kids, from three days to three weeks, need: milk, access to forage and concentrate, water.
- Weaned market kids need good quality forage fed *ad lib*, water, minerals and may need extra protein and energy from concentrate.
- A milking doe needs forages, concentrate, minerals and 6 to 8 liters (1 to 2 gallons) of clean, fresh water. The diet should be 14-16 percent protein.

Having this information, please complete the sentences below:

On my farm I can produce these feeds ______________________________,

______________________________, ______________________________

I have access to these cut-and-carry forages ______________________________

____

I can also provide these grains ______________________________

I must buy ______________________________, ______________________________
(here list things like concentrates)

They will cost ______________________________

I need ____________________, ____________________, ____________________
to produce additional feeds. *(Perhaps it is more land – or additional income – or extra labor)*

Other Comments

____

____

____

____

____

THE LESSON

NUTRITION AND FEEDING

NUTRIENTS REQUIRED IN A GOAT'S DIET

Goats require five major classes of nutrients—water, energy (including fiber), protein, vitamins and minerals.

Water

Goats require from 6 to 8 liters (about 1 to 2 gallons) of water per day. Therefore, it is very important that goats have access to fresh, clean water at all times. Their needs for water are highest during hot weather and lactation as milk is 80 percent to 90 percent water. Although they are able to get much of the water they need from their feed, production will increase if they receive water at least once a day.

Energy

Energy serves as fuel for the animal's body and is thus needed and consumed in the largest quantity among other nutrients. Animals obtain energy from either fats or carbohydrates by a process known as digestion. In this process feeds are broken down into smaller components, such as starch and sugar, which can be absorbed into the animal's blood and used as fuel for movement, growth or lactation. Goats have increased energy needs during periods of growth, breeding, pregnancy and lactation.

While plant cells are made up of sugars and starches that are readily digestible, their cell walls are comprised of complex carbohydrates which can only be digested through fermentation by rumen microorganisms. In balancing rations, the term Total Digestible Nutrients (TDN) is often used as a measure of energy content of a feed. Maize (corn) is an example of a high energy feed. Excess energy is stored in the body as fat, primarily in the goat's abdominal cavity.

Fiber is very important in a goat's diet for proper rumen function. Good quality fiber is the main source of food for healthy rumen microorganisms. Plant material that is 1 cm (3/8 inch) in length or longer is called long coarse fiber. The goat's rumen works best when the fiber-containing roughage rubs against the walls of the rumen, stimulating the muscles to contract and relax—agitating the materials in the rumen. The resulting slurry of materials is more easily digested by the microorganisms that inhabit the goat's rumen. Adequate fiber in a dairy goat's diet also increases the butterfat in the milk. [1]

1) http://www.tennesseemeatgoats.com/articles2/longfiber06.html

Protein (amino acids)

A goat's muscle, hair, hooves, skin and internal organs all contain large amounts of protein. Protein is an important component in enzymes, hormones, antibodies, muscle contractions, oxygen transport and blood coagulation. During digestion, plant proteins are broken down into amino acids, which can be absorbed into the blood and used to make protein for the goat. Unlike energy, protein is not stored in significant amounts in a goat's body. Therefore the animals must receive the required amount of protein for growth, breeding, pregnancy and disease prevention on a daily basis. The amount and type of protein needed varies with activity, including maintenance, production and reproduction. Cottonseed cake, soybean meal, and legume leaves are three examples of protein feeds.

Vitamins

Vitamins are vital nutrients required by goats in very small quantities. Like humans, goats need vitamins A, the B Complex, C, D, E and K. With the exception of D and E, vitamin deficiencies in goats rarely occur because most vitamins are made by bacteria in the rumen during digestion. Vitamin C is made in body tissues. The carotene contained in green, leafy forages is converted into vitamin A. Goats with access to sunlight and/or receiving good quality sun-cured forages will have adequate vitamin D. Vitamin E is found in cereal grains, oil seeds and seed germ. It interacts with selenium. A deficiency of either selenium or vitamin E can cause white-muscle disease in which kids are weak, cannot suckle and may die, while does may have increased reproductive problems. In parts of the world, soils are deficient in selenium. If animals are housed indoors away from the sun, chances of a vitamin D deficiency increase. If goats are fed only dried feed and concentrates, they may also lack appropriate vitamin A levels. [2]

Salt and Other Minerals

Salt and other minerals are important components of a goat's diet for bone development, muscle contraction and production of critical enzymes and hormones for animal well-being. All goats and livestock must have these inorganic elements in small quantities for adequate growth, reproduction and lactation. Balanced nutrition and mineral salt provide needed minerals.

Lactating dairy goats may consume an estimated 7 kg (15 lb) of salt annually, almost twice as much salt as meat goats. A loose salt and mineral mix available at all times is best and should be protected from the rain, but a mineral block may be used as a substitute. In both cases, the mix must be designed for goats, not sheep. If mineral salts for livestock are not available, goats can also be fed iodized table salt.

Goats require calcium and phosphorus at a ratio of two parts calcium to one part phosphorus (2:1). A balance of these two minerals in the feed is especially important for bucks in order to prevent urinary calculi. Moderate to high levels of calcium can be found in most legumes and forbs that goats consume. Lower levels are found in grasses and cereal grains. Some good sources of calcium are: alfalfa, red clover, mulberry, tropical kudzu, seaweed (kelp) and dried citrus pulp.

2) Vitamins and Minerals based on Dr. E.A. B. Oltenacu's Fact Sheet #14 and edited by tatiana Stanton. Used with permission. http://www.abc.cornell.edu/4H/dairygoats/factdg14.pdf

Phosphorus deficiency is of major concern in many parts of the world. In these circumstances a mineral salt containing phosphorus should be used. Other good sources of phosphorus include orchardgrass, sudangrass, rice bran, wheat bran and oats.

WARNING
Ruminant by-products such as bone meal are no longer recommended as a source of calcium.

The mineral content of forages will vary with soil type, rainfall and fertilization practices, so it is important to talk with a local livestock extension person or veterinarian to determine proper feeding practices to provide goats with the needed nutrients, including minerals.

- ***Potassium (K)*** – is found in fresh forages and a supplement is generally not needed.

- ***Iron (Fe) and Copper (Cu)*** – are important ingredients for blood health. A lack of either can lead to anemia. While the small amounts of these minerals present in a goat's regular diet are generally sufficient, some soils are very deficient in copper. Goats grazing on such soils may become anemic and have dull, washed out coats. This can be remedied with the copper found in their trace mineral blocks. Because goats require at least as much copper as cattle, they should be given either cattle mineral or mineral made specifically for goats. Sheep mineral is not sufficient.

- ***Iodine (I)*** – is used by the thyroid gland to produce the hormones that regulate body processes, like metabolism. Iodized salt can be fed to livestock in areas where soil is deficient.

- ***Sulfur (S)*** – is a component of many proteins. Rumen microorganisms need it to build proteins. While most feeds contain sulfur, if using urea or some other non-protein nitrogen source in feed, goats may not receive enough sulfur. (Goats raised for mohair and cashmere fibers need higher levels of sulfur in their diet because hair has high concentrations of sulfur-containing amino acids.)

- ***Magnesium (Mg)*** – is generally found in sufficient quantities in goat feeds. However, lush, fast growing, green pastures that have been heavily fertilized with nitrogen and K can become very deficient in Mg. This can lead to a condition in goats called grass tetany or grass staggers.

- ***Selenium (Se)*** – may be lacking in some soils. Selenium and vitamin E work together to prevent nutritional muscular dystrophy (commonly called white-muscle disease) and retained placentas, and to reduce susceptibility to parasites and disease. Too much selenium can be toxic.

- ***Zinc (Zn), Manganese (Mn), Fluoride (Fl) and Cobalt (Co)*** – are all needed in trace amounts.

MINERAL MIX RECIPE

Combine one part trace mineralized salt or iodized salt and one part oyster shell, dicalcium phosphate or deflorinated rock phosphate.

Recommended Composition of Nutrients as a Percent of Dry Matter in a Mature Goat's Diet:

12 to 16 percent	Crude protein depending on sex, lactation, growth, pregnancy
63 percent	TDN (Total Digestible Nutrients)
16 to 18 percent	Crude fiber
2:1	Calcium-Phosphorus ratio

EXAMPLES OF FEEDS

Energy Feeds
Barley, oats, corn, rice bran, beet pulp, milo, wheat, wheat bran, nuts and roots (sugar beets, turnip, and sweet cassava)

Protein Feeds
Peas, beans, cottonseed cake or meal, soybean meal, linseed meal, coconut, palm nuts, sunflower seeds, sesame, brewer's grains, alfalfa, clover, Erythrina

Calcium Feeds
Citrus pulp, leaves from leguminous plants and leafy green vegetables

Forages: Legumes, Grasses, Forbs, Browse
Alfalfa, Guatemala grass, elephant grass, mulberry leaves, corn stalks, root crops (such as sugar beets, turnip and yucca) and Brassicas (kale, rape, and turnip).

Commercial Feeds
Concentrates, mineral mixes, supplements. (Many farmers grow some or all of their own feed, but sometimes it is more efficient to purchase concentrates or complete rations.)

WARNING
Excess concentrate may cause acidosis and ulcers as well as urinary calculi (stones) in males.

COMPARATIVE FEEDING VALUE OF SELECTED FEEDSTUFFS

FEEDSTUFF	% DRY MATTER	% CRUDE PROTEIN	% DIG PROTEIN	% TDN	% FIBER	% CALCIUM	% PHOSPHORUS
LEGUME HAY							
Alfalfa, avg. analysis	90	15.9	11.7	51	27	1.38	0.20
Alfalfa, early cut	89	19	14.1	59	28	1.41	0.26
Alfalfa, mature	88	13	8.7	50	38	1.18	0.19
Alfalfa, dehy. pellets	92	19	14.1	61	26	1.42	0.25
Peanut (few nuts)	91	7.1	6.1	51	30	1.22	1.01
Cowpea	90	17.5	10.6	53	24	1.22	0.31
Soybean	89	15	10.5	52	35	1.29	0.24
Velvet bean	87	12.9	7.9	56	31	1.10	0.22
Clover, white	90	21	15.9	61	22	1.35	0.32
NON-LEGUME HAY							
Coastal Bermuda	89	10	6.0	56	30	0.47	0.21
Common Bermuda	89	10	6.0	53	30	0.46	0.20
Sudan grass	88	9	5.2	57	36	0.50	0.27
Johnson grass	90	8.6	4.7	48	30	0.75	0.25
Oat, headed	90	10	6.0	54	31	0.40	0.27
Cottonseed hulls	90	5	1.5	45	48	0.15	0.08
Peanut hulls	91	7.0	3.3	22	63	0.20	0.07
GRAINS							
Barley	88	12	7.8	84	5.0	0.06	0.38
Corn, #2 yellow	89	9	5.1	88	2.0	0.02	0.30
Milo or sorghum	89	11	6.9	82	3	0.04	0.32
Oats	89	13	8.7	76	11	0.05	0.41
Wheat	88	14	9.6	88	3	0.05	0.43
Molasses, cane	75	6	2.4	75	0.0	0.97	0.10
Cottonseeds	92	23	17.7	95	29	0.14	0.64
Wheat bran	89	17	12.3	70	11	0.13	1.29
Rice bran	91	14	9.6	72	13	0.07	1.70
PROTEIN FEEDS							
Cottonseed meal	91	48	40	77	13	0.22	1.25
Soybean meal	91	54	45.6	87	3	0.28	0.71
Peanut meal	92	50	42	77	8	0.24	0.58
Linseed meal	91	39	32	76	10	0.43	0.93
Bean, pinto	90	22.6	17.2	79	4.0	0.13	0.46
Cowpea seed	89	23.8	19.5	76	5.0	0.09	0.44
PASTURE/RANGE							
Alfalfa, avg.	24	19	14.1	61	27	1.35	0.27
Coastal Bermuda	29	4.4	3.4	21	8.8	0.14	0.08
Common Bermuda	34	4.1	unknown	23	10.5	0.18	0.07
Honeysuckle	36	10	5.9	69	9.4	0.40	0.11
Oat, immature	16	2.5	unknown	10	4.0	unknown	unknown
Ryegrass	24	3.8	1.6	15	6.8	0.17	0.11
Sudangrass, avg.	18	17	12.3	70	23	0.46	0.36

SOURCE: NRC REQUIREMENTS FOR SMALL RUMINANTS 2006

FEEDING GOATS DURING WET, DRY AND COLD SEASONS

Wet Season

Pastures can be managed for maximum production during the rainy season through rotation, vegetation height maintenance, adjusting livestock density and resting the pasture. It may be necessary to limit browsing during the rainy months. Shelter animals during heavy rains and allow them to browse when it is not raining. If rains persist or animals are not allowed to browse long enough to meet their nutritional needs, be prepared to take forage to the goats.

WARNING
Parasites thrive on moist grasses and can easily infest animals.

Dry Season

At the beginning of the dry season, forage will be plentiful. This is probably the best time for kids to be born and to develop your milking animals. As the season grows longer, there will be fewer bushes and grasses for the animals to browse. For this reason, cut and store hay early in the dry season to use when it is no longer plentiful. Another suggestion is to plant forage trees like Leucaena or Acacias that will provide feed even during drought.

Cold Season

During severe cold weather, it is recommended that goats be brought into shelters or barns to keep warm. Good ventilation keeps the manure and bedding pack dry. It also keeps ammonia out, while reducing respiratory problems and skin infections. (Moisture on barn walls is an indication of inadequate ventilation.) Preserved feeds such as hay or silage can be fed during cold weather along with concentrates. Ensure adequate vitamin intake by using high quality feeds. As weather permits, the goats should spend some time outdoors to exercise and acquire vitamin D from the sunlight.

FEEDING NEWBORNS, ORPHANS AND YOUNG KIDS

Colostrum, the thick and often pale yellow substance produced by the doe for 72 hours after parturition (kidding), should be consumed by the kid within the first hour after birth. It is in fact the only food a newborn kid should consume. The doe will continue to produce colostrum for 2 to 3 days. Colostrum obtained from the first milking after kidding contains more protein (especially antibodies), fat, minerals and vitamins than milk produced in later lactation. Antibodies are proteins produced by the doe's immune system that protect the kid from infectious disease. They are secreted into the colostrum around the time of kidding. The sooner a kid gets colostrum the more antibodies will be absorbed.

If the doe should die or be unable to produce colostrum or enough milk for her kids, a substitute will be needed. Milk is not a substitute for colostrum. Ideally, the colostrum from another doe is the best substitute for at least the first day or two of a newborn's life. After that it may be practical to use goat milk from other does. However, it may be more cost effective to use powdered milk substitutes made especially for goats. If these are not available, use powdered milk or cow's milk.

In the early stages, consider feeding smaller quantities more often until the kid's digestive system adjusts.

First Week

Feed a kid as much colostrum as it will consume in the first 12 hours of life. Leave kids with doe, or allow to nurse six times a day for three days. After three days, most kids are ready to be separated at night. The doe can then be milked in the morning for human consumption. Kids can remain with the doe during the day whether zero grazed or pastured. Kids should have access to hay and roughage to develop the rumen.

Second Week

Offer the kid up to ½ liter (about 1 pint or 2 cups) of milk at least three times each day. Put protein concentrate, hay and water near the kids and they will learn to eat and drink well. Use creep feeder if housed with adult goats. Be sure the kids cannot fall into the water and drown.

4 cups = approximately 1 liter

Third to Sixth Week

Give 1 to 2 liters of milk divided into three feedings, along with concentrate, hay and water. The rumen in kids begins functioning at about 21 days of age.

Seventh to Ninth Week

Decrease number of milk feedings to two a day. Continue to feed concentrate and forage.

Ninth to Twelfth Week

Decrease milk slowly to once a day. Continue with concentrate and hay or forage. Wean kid. If kids are left with does all the time, they usually wean themselves at six to 12 weeks. Although kids may still want to nurse after 12 weeks, milk is no longer a nutritional requirement and the rumen should be developed. It is important, however, that kids continue growing. Weigh the kids every week and record their weight. In North America and Europe, kids from large-sized dairy goat breeds will weigh about 15 to 20 kilograms (30 to 45 lbs) at the weaning age of three months. However, in developing countries the healthy weight of local breeds may be lower.

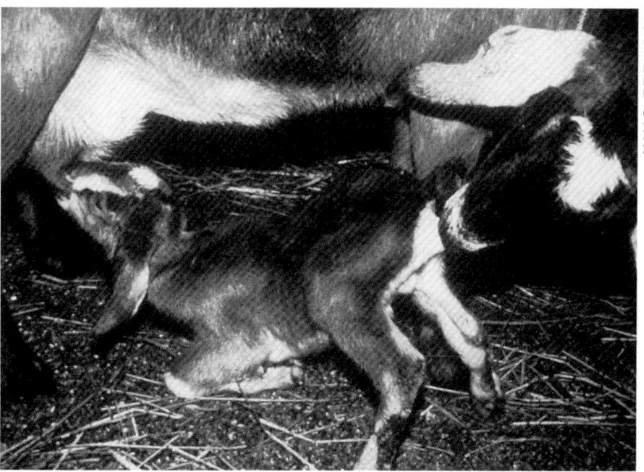

ROSALEE SINN

Milk for Family Use While Kids are Nursing

A doe may be milked for human consumption without jeopardizing the nutritional needs of her kids. If kids are left with the doe until weaning, the best way to guarantee an extra supply of goat milk is to take the kids from the doe each evening. Place three to four kids in a 1 x 1 meter (3 x 3 foot) enclosure with solid sides to prevent unhealthy drafts. Each morning, milk the doe and reserve it for consumption or sale. Release the kids from the box to spend the day with the doe and feed freely. Repeat this procedure each day.

Providing Forages and Concentrate to Kids

Kids need forages like grass and legumes to grow properly and have a well-rounded diet. This is important for adequate rumen development before weaning. Be sure the kid gets plenty of good mixed hay, browse or pasture, a protein concentrate, mineralized salt and water.

If the doe is penned and the kids are left with the doe, it is best to make a creep feeder for the kids. This will allow kids to get forage, water and feed, without competition from the doe.

Hand-feeding Kids

Bottle Feeding

A baby bottle or soda bottle with a nipple attached can be used for bottle feeding kids. Make the hole in the nipple large enough that some milk dribbles when the bottle is turned upside down. Be sure the bottle and nipple are clean at every feeding.

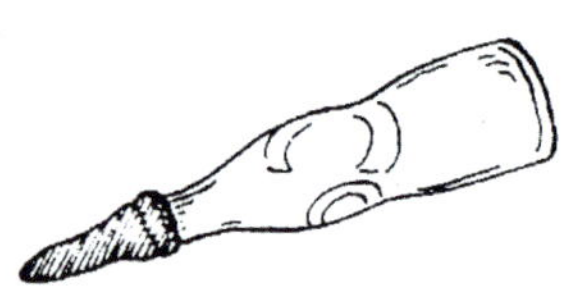

Pan Feeding

A kid can be taught to drink from the pan. Do so by placing milk in a pan, sticking a finger in the milk and letting the kid suck from the finger while putting its mouth into the milk. Pan feeding is often easier. Wash the pan with soap and water and sun dry after every feeding.

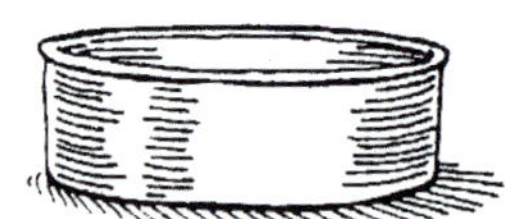

Bucket Nurser

As the goat herd grows and there are many kids, a bucket nurser may be a valuable tool as up to eight kids can be fed at a time. A bucket feeder consists of a plastic bucket, some kind of plastic tubes and nipples. Note that it is very important that the tubes, nipples and bucket be kept clean. Place bucket at the correct height on a sturdy holder, which can be made from a wood base and four boards for corners. Slow nursers may need special attention. Smaller or weaker goat kids may be repeatedly crowded off of their nipple and may not get their full quota of milk.

Lambar Nipples

Australian-made natural rubber nipples for Lambar style feeders are available in local feed stores or can be ordered from: Caprine Supply, P. O. Box Y Desoto, KS 66018 U.S.A. or via internet at *www.CaprineSupply.com* This nipple fits through 5/8" (1 cm) hole which should be placed 7" (18 cm) above bottom of bucket. The free end of the tubing extends to the bottom of the bucket.

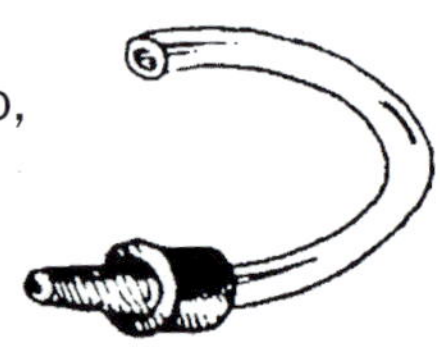

Multiple Bottle Nurser
This is a homemade wooden box nurser to hold bottles with nipples.

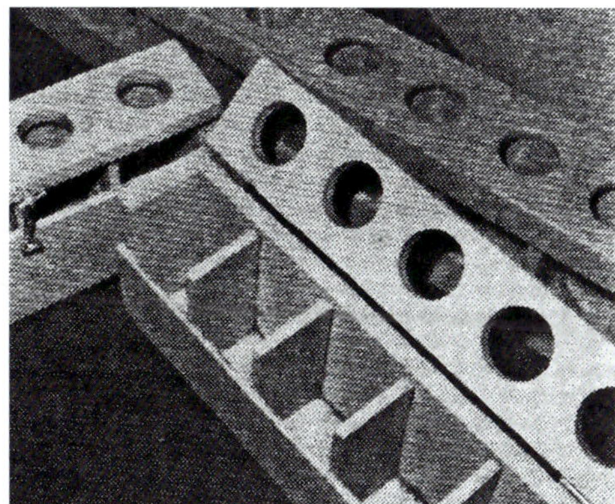

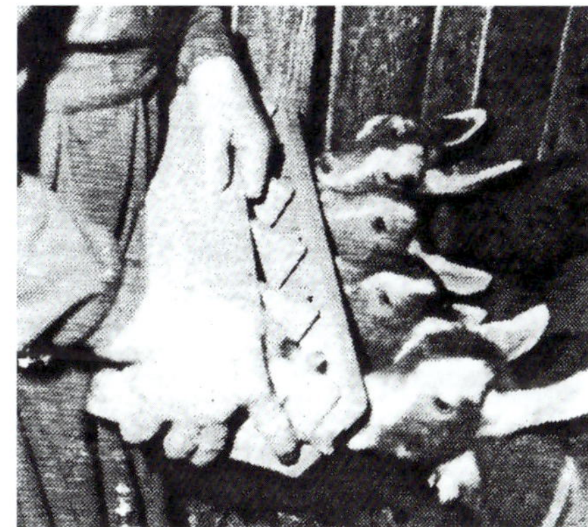

BENEDICTINE NUNS, MONT-LAURIER, QUEBEC, CANADA

CAPRINE ARTHRITIS ENCEPHALITIS (CAE) PREVENTION PROGRAM

If CAE is a problem in the surrounding area, consider adopting this program practiced by goat farmers. Before beginning a CAE prevention program, discuss it with your veterinarian, extension person or with another breeder who manages goats in this manner.

AT BIRTH

Take kids from doe immediately. Do not let kid suckle. Do not let doe lick kids. Towel dry kid. Feed kid(s) at least 2 ounces (oz) heat-treated colostrum within one hour of birth. Continue to feed colostrum at four-hour intervals during first day.

FIRST WEEK

Give ½ liter (1 pint or 2 cups) pasteurized milk or milk substitute 4 to 6 times each day.

SECOND THROUGH TWELFTH WEEK

Follow feeding schedule in normal program using pasteurized milk or milk substitute. Keep kids away from physical contact with adult goats.

HOW TO HEAT TREAT COLOSTRUM

If possible, have heat-treated colostrum from another doe available for immediate use. Cow colostrum may also be used, but only if no goat colostrum is available. Fill stainless steel thermos with water that has been heated to 135°F. Seal for 1 hour. The water should be at 131°F or a few degrees higher when removed from thermos. Heat colostrum to 131°F. Watch carefully. Do not exceed 131°F or you will destroy antibodies. Pour water out of thermos. Pour in heated colostrum. Seal for one hour. Refrigerate or freeze in another container for future use.

FEEDING MATURE GOATS

Always give a goat more green chop than they can eat. If cutting and carrying the forage to the animal, place it in a manger or use a tie to keep it off the ground. Feeding forage in this way will reduce waste and decrease parasite ingestion. Goats should be allowed to eat as much forage as they desire.

> **WARNING**
> **An obese doe is more likely have trouble kidding and will be more susceptible to pregnancy toxemia and fatty liver syndrome. Obesity in animals causes lack of energy utilization and may also impact the ability to be bred.**

Pregnant Does

Stop milking dairy does when they are three months pregnant. Any kids that are still nursing should be weaned at this time. If a doe has been fed high-calcium legumes (clover or alfalfa), it is best to gradually replace these with grass hay at least three weeks prior to kidding in order to prevent milk fever. (See Health Chart in Lesson 8). This forces the animal to mobilize her body's own calcium reserves and prepare for lactation.

After three months, the kids inside the doe will be growing fast. During these final weeks, the unborn kids will require the most nutrition of the entire pregnancy. Be sure to make more feed available to the doe during this time, along with mineralized salt and water. She may also need some concentrate to provide enough energy and protein depending on her body condition, the number of kids she is likely carrying and the quality of the forage.

Lactating Does

A 50 kilogram (110 lbs) dairy goat that is milking will be most productive on a diet of green chop (forage) and at least 0.5 kg (1 lb) of concentrate each day containing 14 to 16 percent crude protein. Feeding requirements are greatly increased during lactation. If you feed a prepared concentrate, feed 0.5 kg of concentrate for each 1.5 kg of milk produced. High quality hay will provide extra protein. A mixture of green chop and legume hay, which is particularly high in protein, is excellent for dairy animals. Legume hay is made from special grasses and high-protein plants, such as alfalfa (lucerne), desmodium spp. (silverleaf desmodium, greenleaf desmodium), clover, beans, peas and groundnuts. Alfalfa is particularly good hay for milking goats because it is high in calcium.

Milk is 80 to 90 percent water. Remember that adequate water is one of the most critical needs for the milking doe. Be sure a milking doe receives 6 to 8 liters of clean water each day. If the water is contaminated by feces or other foreign matter the goat will not drink it. In cold weather, goats prefer warm water, if available.

> **WARNING**
> **Do not feed a milking dairy goat onions, garlic or other forages that will give the milk an unpleasant taste. Keep mature bucks separated from the milking does since their odor can also cause the milk to have a bad taste.**

Breeding Bucks

Proper feeding is very important for breeding bucks. Large quantities of high quality hay or fresh forage will help maintain and grow bucks at a reasonable cost. Plenty of clean water, a proper calcium/ phosphorus balance and exercise are also important for the breeding buck's health. As the breeding season approaches, concentrate or other energy and protein concentrate can be gradually added to the buck's diet until reaching 0.5 kg (1 lb) a day (for a 60 kg or 132 lb buck). As a buck needs only 12 percent protein in his diet, after the breeding season, concentrate feeding may be discontinued. When not being used for breeding, be sure the buck gets plenty of green browse and pasture.

> **WARNING**
> **Never let a buck get fat or sluggish. This may cause sterility. High temperatures or fever from infection can also cause temporary sterility.**

Meat Goats

Meat goats can be fed mostly forages to meet their nutritional needs, unless the market for fat goats is strong. Adequate forage will provide energy, minerals and vitamins in sufficient dietary amounts. Since goats are particularly adept at selecting the most nutritious plants (and within plants, the most nutritious portions), they do reasonably well grazing areas that are unsuitable for cows, provided the amount of herbage is adequate. Like other animals, goats respond quite favorably to increased quality in their feeds. Water requirements of meat goats vary, but having a good source of water is important.

Trimmings from a vegetable garden such as beet greens, dried cassava leaves, cabbage leaves and bean vines are also good sources of feed. Goats will also eat beans, groundnuts or bambara groundnuts, sunflower seeds, banana peels and various types of fruits and vegetable peelings.

> **WARNING**
> **Do not feed too many banana peels as this can cause diarrhea. Avoid green potatoes and tomato and potato vines as they contain solanin, which is toxic.**

EXAMPLES OF FEEDS USING OATS AND WHEAT BRAN

	POUNDS	KILOGRAMS
Ration for a milking doe fed good legume hay - 13 percent digestible protein		
Corn	31	14
Oats	25	11
Wheat Bran	11	5
Linseed oil meal	22	10
Cane Molasses	10	4.5
Salt	1	0.5
Ration for a milking doe fed good legume hay - 13 percent digestible protein		
Barley	40	18
Oats	28	13
Wheat bran	10	4.5
Soybean oil meal	11	5
Cane molasses	10	4.5
Salt	1	0.1
Ration for a milking doe fed non-legume hay - 21 percent digestible protein		
Corn	11	5
Oats	10	4.5
Wheat bran	10	4.5
Corn gluten feed	30	14
Soybean oil meal	11	5
Cane Molasses	10	4.5
Salt	1	0.5
Ration for a milking doe fed non-legume hay - 21 percent digestible protein		
Barley	25	11
Wheat bran	10	4.5
Soybean oil meal	25	11
Linseed oil meal	15	7
Salt	1	0.5
Ration for dry does and bucks - 10 percent digestible protein		
Corn	58	26
Oats	25	11
Wheat bran	11	5
Soybean oil meal	5	2
Salt	1	0.5
Ration for dry does and bucks - 10 percent digestible protein		
Barley or wheat	52	23
Oats	35	16
Wheat Bran	13	6
Salt	1	0.5

SOURCE: ROBERT A. VANDERHOOF, *RAISING HEALTHY GOATS*, CHRISTIAN VETERINARY MISSION, 2006.

Goats Provide Meat, Income and Self-Esteem

When Heifer International started working with women in Nepal, the women were quick to identify the niche market for goat meat. Considered a delicacy, goat meat sells at a very good price in Nepal. As compared to NPR 80 (NPR 75 at 1 US $) per kg of buffalo meat, NPR 130 per kg of chicken and NPR 140 per kg of pork, goat meat is the most expensive, priced at NPR 250 per kg. There are strong market demands for male goats during Hindu festivals like Dashain. There are 38,584 metric tons of goat meat produced annually in Nepal, second to 127,495 metric tons of beef. It is followed by the production of 35,000 metric tons of fish, 15,594 metric tons of pork and 14,399 metric tons of poultry.

▲ *Passing on the gift in Nepal.*

Eighteen of the 23 Heifer Nepal projects involve goats. It is also important that most of the members of these groups are women. Since December 2002, the 3857 goats given through Heifer International have helped 2114 Nepali families. Goats are cheaper to buy and occupy less space than cattle. Goats can be fed a wide variety of forage. They do not need specially formulated feed and can eat kitchen leftovers to meet their dietary requirements.

In addition, they can be bred at seven to eight months and with a 150 day reproduction cycle, they are quick to produce income. This makes goats attractive to small-scale farmers. Many goat farms in Nepal breed does three times in two years, making it possible for a single female goat to produce seven to nine goats in two years. Adult male goats are sold at about 18 months, when they yield at least 20 kilograms (44.09 lb) of meat. Male kids are sold just after weaning.

Goat projects have been highly successful and stories from the field are very encouraging. The story of Ganga Devi Khanal's struggle, determination and perseverance is one of many. Despite being dissuaded by her husband and family, she never gave up. Ganga's husband, who originally was against his wife's efforts had even dragged her away

Continue on page 48

Continued from page 47

from group meetings.

She had very little knowledge of animal management and agriculture. She could not read and write or speak in front of others. She did not know about proper nutrition and how it affected well-being. Their mud thatched hut and the animal shed were very unclean and ill-kept. She had to live with insufficiency, agony and anguish. It was very frustrating but she continuously looked for ways to come out of this painful life.

When she met the women group members of Heifer Nepal project in Gitanagar and learned about their activities, she was inspired to form a group. In August of 2000, the Women's Coordination Committee, a local NGO working with Heifer Nepal, helped her to establish a group named "Nari Uthan Women Group" consisting of 20 Tharu and Brahmin women. She became the president of the group. The group started saving money, initially with NPR 20 per month and later increased to NPR 50 per month. With the fear of rejection from the family she kept quiet and did not let her family members know about the group activities for the first two months.

In 2001 Ganga Devi Khanal received two goats along with training on the Heifer International Cornerstones, group management and livestock and fodder management. A new hope for a prosperous life was kindled. This gave her strength to work harder.

She earned NPR 35,000.00 (About US$466) from the goats. Part of this income was used to improve her family's living conditions, to provide enough nutritious food and to pay for her children's education. With NPR 14,000.00, Ganga bought a buffalo, which gives 6 liters of milk a day. She earns NPR 14 per liter from the sale of excess milk. Use of manure compost has increased the fertility of her fields and increased the yield of crops and vegetables.

At present, Ganga has a buffalo, a cow and five goats. She has also been able to construct a concrete house, a toilet and an improved shed for animals. Ganga first passed on two goats. Later, the dignity of being a donor encouraged her to pass on one more.

Ganga Devi Khanal was proudly sharing her success stories with Heifer Nepal team. Suddenly her eyes filled with tears and her voice choked, when she saw her husband serving tea to the guests. Wiping her tears with one hand she remarked "It is so satisfying to see him helping out and supporting me in everything I do." With beaming eyes, Ganga remarked, "All the pride, wealth and respect I have earned today are because of the goats. I vow to do my part by helping as many people as I can." ◆

THE LEARNING GUIDE

FORAGE PRODUCTION, PASTURES AND ENVIRONMENTAL MANAGEMENT

LEARNING OBJECTIVES

By the end of the session, participants will be able to:

- Describe the relationship between goats and the local environment
- Identify grasses, browse, forbs and nitrogen-fixing trees in local area
- Share a basic understanding of the feeding behavior of goats, forage production and pasture improvement

TERMS TO KNOW

- Forage
- Forbs
- Browse
- Nitrogen Fixing Trees
- Legume
- Agroforestry
- Fertilizer
- Compost
- Biodiversity
- Silage

MATERIALS

- Pads and pencils
- Cut and labeled forages
- Cut and labeled poisonous plants
- Machetes or cutting knives
- Bags or baskets to gather forage

ADVANCE ASSIGNMENT

- Define areas for forage activities
- Cut and label forages from the region
- Ask the class to bring machetes or cutting knives
- Acquire bags or baskets for gathering forage

TIME (May vary according to group)	ACTIVITIES
20 minutes	**Group Sharing** Do you have any questions about last week's session? What changes did you make? What do you hope to learn from this session?
45 minutes	**Get Everyone Thinking and Talking** ■ Give the group pads and pencils. ■ Ask the group to walk in teams around the farm or compound, observing livestock, especially goats. Encourage them to make notes about how the goats are interacting with the environment, what the goats are eating, which part of the plant they are eating, etc. ■ Ask each team to report their observations and to share any questions that arose during their walk.
15 minutes	**Identify Samples of Forages** ■ Browse, forbs, grasses and trees suitable and not suitable for feeding goats. Identify common poisonous plants in the area. ■ Have a local farmer tell about how she/he uses the various forages. Discuss how they help balance the goat's diet.
60 minutes	**Cutting Local Forages.** ■ Send teams to defined areas to cut forages and bring back to group. ■ When they return, discuss: ■ Whether to feed fresh or dry and why (i.e. some plants are poisonous until they are dry, such as bitter cassava) ■ The stage of growth at which to cut various forages.

REVIEW-20 MINUTES

- What was useful?
- What was a surprise?
- What did not work out well?
- What do you know now that you did not know before?
- What can you do as soon as you get home?
- What practices will be difficult to do at home?

THE LESSON

FORAGE PRODUCTION, PASTURES AND ENVIRONMENTAL MANAGEMENT

Goats are energetic, inquisitive and versatile in the art of gathering food. They can consume large amounts of browse (tree leaves, bushes, twigs, etc.) in contrast to cattle and sheep that prefer grasses and legumes. They learn their eating habits from the herd. Thus, goats in one community may not eat exactly the same types of feed as goats in a different community.

Their mobile upper lip enables them to discriminately select favored parts of plants. Goats chew their food more completely than cattle and therefore digest their food more efficiently. These two traits can sometimes allow them to obtain a higher per cent of digestible material from their diet than other animals. However, a goat is a small ruminant which means that food travels through its digestive tract faster than it does through a large ruminant. Therefore, they have less ability to ferment and digest extremely coarse fiber compared to large ruminants.

INTRODUCTION TO FORAGES

Forages are the vegetative parts of the plants that goats eat, and should form the bulk of a goat's diet. Unlike humans, the digestive tract of the goat is able to use the cellulose found in forages for the majority of its energy needs. Forage plants for goats can be classified into four main groups:

- browse
- forbs
- legumes
- grass and grass-like plants

Note: There is some overlap between these groups such as "leguminous forbs".

Browse

Browse refers to plants with woody stems such as shrubs and small trees. Goats prefer the leaves and stem tips of many browse plants when available. Although browse plants leaf out rapidly at the end of winter or at the beginning of the wet season, they do not have the ability to regrow their leaves quickly after grazing. Thus they can only be grazed once or twice during a growing season unless the goal is to eradicate them. Their protein content decreases slightly as the plant matures over a growing season.

Forbs

Forbs are broadleaf, herbaceous (non-woody) plant species. Some forbs may have somewhat woody stems but unlike browse plants these stems die back every year. Forbs include a number of plants that farmers often consider weeds, but which are actually very nutritious for goats (for example, chicory, dock, and various members of the Brassica family). Forbs generally grow back fairly rapidly after grazing until they mature enough to flower and make seed. Similar to browse, their protein content decreases slightly with maturity.

Legumes

Legumes are members of the "pea family," a large and widespread family of plants, including alfalfa, acacia, clovers and leucaena and legume fruits such as beans, peas, and soybeans. A distinctive feature of the family is the presence of root nodules containing rhizobium bacteria, which convert atmospheric nitrogen so the plant does not depend on nitrogen from the soil or a fertilizer to grow or make protein. Because of the rhizobium-containing nodules, legumes actually contribute extra nitrogen to other forbs and grasses in the same pasture, allowing these plants to grow better. Pasture legumes make pasture grasses more productive.

In many cases, the pods of edible legumes are also nutritious, though sometimes the pods and leaves contain toxins. Tree legumes are often called nitrogen-fixing trees (NFT). Leguminous forbs in pastures generally grow back rapidly after grazing. Depending on water availability, they may be grazed repeatedly during a growing season, providing a nutritious, high protein feed.

Grasses

Grasses are herbaceous plants with round hollow stems and flat narrow blades for leaves. Grasses harvested at the right stage are high in energy and contain moderate amounts of protein. Upon maturity, their protein content drops substantially more than that of forbs or browse. Grasses are low in calcium and animals grazing them may need a calcium supplement such as calcium carbonate (lime) in their mineral mix.

If rainfall is adequate and temperatures are not too warm or cold, grasses will grow back rapidly after grazing or cutting. Their yields improve if they are planted together with legumes or improved with manure or chemical fertilizer.

GOATS AND ENVIRONMENTAL MANAGEMENT

Goats can damage fragile environments if not properly managed by their keepers. If stocking rates are too high for the land in question or goats are not controlled during the dry season, overgrazing can occur. This results in long-term damage and erosion. As goats readily girdle (remove bark from) shrubs and young trees, they can cause substantial damage to orchards, tree plantations, or fragile shrub/forest ecosystems. All of the these scenarios, however, result from inadequate goat management.

When properly managed, goats can serve as highly effective tools for improving the environment. The small cloven hooves of goats cause little damage except on soils where grazing should not be permitted by any livestock species. Compared to other farm species, goats are excellent browsers (see table below). This means that they will select much of their diet from the leaves and young stems of woody plants when available. These plant parts provide more energy, protein and minerals than the rest of the plant.

DIET PREFERENCE DIFFERENCES PERCENT OF DIET			
Species	**Grass**	**Forbs**	**Browse**
Horse	81	5	14
Cattle	70	18	12
Sheep	57	23	20
Goats	39	10	51

SOURCE: VALLENTINE, J.F. (2001) *GRAZING MANAGEMENT*, ACADEMIC PRESS

When goats eat browse, they help eradicate brush and re-establish savannahs and grazing lands and can rejuvenate abandoned hayfields or croplands. In wooded areas, they can reduce fire hazards by maintaining fire breaks and enhancing timber production by clearing excessive understory.

Goats can also be used to control or clear undesirable species. Goats will preferentially graze many invasive plant species such as American kudzu and Japanese honeysuckle. By controlling these unwanted species, goats can help an area regain biodiversity and re-establish native plants. Their readiness to consume many species of browse and invasive plants, makes them ideal for cleaning irrigation ditches and farm ponds, reclaiming fence lines, creating flyways for waterfowl and keeping brush from encroaching on community commons and other lands.

Grazing Management

Grazing management can be defined as the movement of livestock through pastures and shrubby areas to provide nutrition for the livestock while enhancing the vegetation and soil. Farmers with little land and only a small herd of goats may not have many grazing options, but when access to idle lands or shrubby areas is available the following strategies may be useful.

The forage base, whether grazed or cut and carried, is the most important part of a goat's diet and needs to be managed properly to remain productive. If goats are left in the same pastures all the time their ability to selectively graze will allow them to wipe out the most desirable plant species while letting undesirable species multiply. Therefore, it is important to keep the goats moving through areas of the farm, forest or pasture, not allowing them to stay very long in any one area.

This can be most easily accomplished by fencing off small areas within a larger pasture or browsing area. To get started with rotational grazing, current pastures can be divided in halves or quarters. When one area is grazed to approximately 15 cm (6 inches), and no more than half the leaf area of browse plants is left, move to the next area. Intake may decrease as goats are forced to graze lower than 6 inches. Pastures should never be grazed lower than 3 inches. As pastures are grazed and given adequate time to rest, plant diversity will change and improve. Less desirable plants will usually be crowded out and more desirable plants become a larger portion of the pasture. Goats will have more of their preferred forages to eat, which improves their health and productivity. Use a scythe or machete to eliminate unwanted, useless or toxic plants.

Ideally goats will be grazed on a new pasture area while the forage is still relatively immature to help delay flowering of the plants. Most of the available protein in plants is found in the leaf and young stem tips. Grasses, forbs and leguminous plants are all highest in protein prior to flowering. After flowering, the protein content of these plants drops and the cell walls are less digestible and less nutritious. A pasture with a selection of grasses, legumes, forbs and browse is ideal for goats.

Community grazing systems are most functional in areas that have community-owned land with large amounts of crop residue, such as from sugar cane. The benefit of community grazing is that all members of that community can share forage resources, while clearing crop residue and treating the soil with goat manure. However it is important that each goat be in good health in order to prevent transmission of parasites and disease between farms.

Forage Management

Pastures and other areas used for growing forages can be managed to enhance production and to improve the environment. Soil fertility, plant species, and growing conditions will affect the yield and nutritional value of forages whether grazed or harvested. The productivity of a grazing area is also impacted by animal density, the use of compost or fertilizer, the stage of maturity when grazed, the grazing time and the amount of re-growth allowed before re-grazing.

TATIANA LUISA STANTON

▲ *Goats foraging.*

PLANT MANAGEMENT AND SOIL IMPROVEMENT

Before establishing or attempting to improve a pasture, request a soil analysis to identify the current soil conditions. It is important to know what nutrients are present and which are lacking from your soil in order to make improvements that will guarantee the growth of nutritious forage. Any missing nutrients may be replaced through the application of manure, compost or chemical fertilizer.

Properly fertilized pastures are much more productive and generally less susceptible to drought and other stresses. If soil is too acidic, add lime (calcium carbonate) or choose a plant species adapted to acidic soils. Adding organic matter to pasture soils improves the soil's ability to hold water and nutrients.

Establishing New Plant Species

Check with local farmers and extension staff to find out what nutritious goat forages are easily grown in the local area. The profile of an existing pasture can sometimes be modified by a process known as over-seeding. In this process, the seeds of a more desirable forage species, like legumes, are added to the grasses. First, graze a pasture low, sow the new seed and then concentrate a large number of goats or other livestock on the area to help drive the seed into the ground before it rains.

In some cases living fence posts or hedgerows of nutritious trees and shrubs can be planted on pasture boundaries to provide extra nutrition. When planting a tree to be used as forage during the dry season, choose a species that does not shed its leaves during this season. Protect young trees for about two to three years from grazing goats. Old tires or cement blocks serve as excellent barriers during the early growth of these trees.

Forage diversity and pasture productivity can be increased by planting or seeding both grasses and legumes. Less useful plants can be discouraged by repeatedly cutting or mowing. Improved pastures can be planted under crops such as fruit or timber trees to increase the total productivity of land.

ERWIN KINSEY

▲ *Forage in Contour Plantings in Tanzania.*

Forages can be used as contour barriers, green manure crops and bank protection along ponds and waterways. In these environments, harvest only as much forage as these plants can safely sacrifice while still fulfilling their conservation role. Grass and legume contour barriers will protect the soil and can be cut periodically to be fed to livestock in zero-grazing and semi-intensive systems.

PARASITE CONTROL THROUGH FORAGE MANAGEMENT

Internal parasites, especially stomach intestinal worms, are the biggest disease problem for goats. Practicing pasture rotation with long rests between reuse of the pasture, increasing the availability of browse and implementing other practices mentioned below can help the farmer to be more successful with fewer costs and healthier goats.

If parasite free, goats could be released to graze on a field just as soon as the forage recovers. This so-called "intensive grazing" is difficult because most pastures are not worm-free until two or three months after grazing. With such long rests, pastures often reach maturity before they can be grazed again and are not as productive as they could be. To prevent high worm loads as a result of repeatedly grazing one area, move goats to browse other areas, feed them crop residues or put them in a zero grazing situation for a few months.

Deworming goats at the beginning of the grazing season may aid in cutting internal parasite numbers in the pasture. Grazing cattle, horses or donkeys on the pasture before the goats return also reduces the number of worm eggs, allowing goats to safely return to the pasture sooner. Keep in mind that weaned kids, lactating does and pregnant goats are the most susceptible to parasites and should have first access to the cleanest pastures. Mowing a pasture low and harvesting it for hay in very dry, hot weather will also kill worm eggs.

Browse grows taller than most forbs and grasses. As the larva of most parasitic worms cannot climb higher than two inches on plants, goats are less likely to re-infect themselves with worm eggs when eating browse compared to forbs or grass. Many browse plants contain high levels of bitter substances called tannins, which seem to reduce internal parasite loads.

Manure and Parasites

Manure dropped during grazing or used as fertilizer on crops can contaminate an established pasture with internal parasites (worms). Chicken, cow, donkey, horse or pig manure will not spread worms to goats. Similarly, the worms found in goat manure will not infect other species. However, the worm eggs from sheep and goat manure can infect both species.

To prevent the spread of worms, it is best to compost manure first or spread it on the field several months before any grazing takes place. If goats are confined all day or brought in at night, their manure and bedding can easily be collected and turned into compost. Make a large pit to store the manure and bedding away from where humans cook or sleep since it will have a strong smell in the beginning. Do not add meat, fat or human waste because this may spread disease. Add plant material or food scraps and allow to heat up for several weeks. Turn the pile often and add water.

In a short time (a few weeks in hot weather and a few months in cold weather), the material will become a black and crumbly compost with a mild acidic odor. Use this compost on your pasture or on other crops. Compost is better than store-bought or “chemical” fertilizer because it contains organic matter which allows the soil to retain more moisture and nutrients.

Toxic Plants and Lush Pasture

Check with other farmers and livestock extension agents to find out what local plants are toxic to goats. Yew, oleander and rhododendron (azaleas, etc.) are all common ornamentals or landscaping plants that are very toxic to goats. Unfortunately goats are very eager to eat them when other nutritious feeds are scarce.

Certain plants such as trees and shrubs in the cherry family, bitter cassava, sudangrass and johnsongrass can be poisonous to goats. The leaves can be fed when wilted, during or after a drought or after mild or severe frost. Under these conditions, prussic acid is released causing toxicity, especially in immature, growing grass. Once completely dry, the grass is safe.

Feeding too much lush pasture too suddenly can increase the likelihood of diarrhea, enterotoxaemia and/or bloat. Always remember to change a goat’s diet slowly and offer a range of forages to help cut down on accidental poisoning.

CUTTING AND STORING FORAGE

Hay

If shelter is available, forage can be stored for other seasons when it is not readily available. Planning ahead for these times is important. Ideally, grasses should be cut for hay prior to flowering and leguminous forbs should be cut in the early bloom stage (just as they begin flowering) to assure high digestibility and high nutritional content. However, more mature plants will be taller and easier to cut (if cutting by hand) and will also yield more dry matter. Often one needs to make a choice between greater yield or greater nutrition. Delaying too long greatly increases the indigestibility of grasses and legumes and substantially decreases their protein content.

Cut hay (forage), dry it and bring it to a barn or shelter to store for feed during the rainy season, in winter or during the extreme part of the dry season. Try to make hay during rain-free weather. Leave hay to dry one or two days before baling or stacking. Turn the cut hay once or twice each day so it dries well. Hay can then be stored for many weeks, providing good feed when grazing is not possible. Elephant grass grows well in many places and is easy to cut by hand for hay. Sudangrass and Guinea grass (buffalo grass) are also good choices for hand cutting. Shorter growing grasses (pangola, timothy, etc) and leguminous forbs (alfalfa, clover, etc.) produce more nutritious hay, but are far harder to cut by hand.

WARNING
Hay that contains too much moisture or large stems will mold and be unsuitable for feeding. Closely examine and smell all hay before feeding it. It should always smell sweet.

Silage and Haylage Production

Silage is fermented plant material with an acidity level that prevents spoilage. Although less palatable to goats than hay, silage has storage advantages. While larger amounts of silage may be stored in a trench silo, small amounts of silage may be prepared and stored in plastic bags. Silage must be kept in an air-tight container until fed. After cutting and chopping the forage, press tightly into the bag. Press out all of the air and tie the opening to form an air-tight container. Before using as feed, examine closely and smell. Good silage smells sweet and slightly acidic, never rotten.

Corn silage is an acceptable feed, but is often fed with other forages and concentrate. Corn silage is made by chopping the stalk and the ear.

Making haylage out of harvested legumes and grasses is an alternate option for storing and processing forage. Haylage can be made from any crop that is traditionally stored as hay.

Ensiling grasses and legumes as haylage requires 40 to 60 percent moisture. Reductions in moisture content necessary for production of haylage are accomplished by conditioning (mowing, windrowing and drying for four to 24 hours). The drying time depends on moisture content and weather conditions. As with silage, the forage must be maintained at a very acid pH (4.5) to prevent spoilage. A livestock extension agent can advise on the proper techniques for making silage or haylage in your area. For silage of legumes

WARNING
Feeding poorly prepared or stored haylage or silage can lead to listeriosis or circling disease if the organism is present during processing. Listeriosis is almost always a danger when making either haylage or silage.

and other plants with high protein content, it is essential to add molasses (5 percent) or another high energy ingredient to enhance the fermentation process.

A SAMPLING OF COMMON FORAGE PLANTS FOR GOATS

The following pages contain descriptions of various grasses, forbs and legumes, including nitrogen-fixing trees. Most of the forages described in this section are from tropical areas of the world. Consider adding local forages to this list, along with information on their nutritional value and how they can best be grown in local conditions.

GRASSES

African Star Grass–*Cynodon aethiopicus*

(Rhodesian star grass; African bermuda grass; Giant star grass)
African star grass is a hardy, perennial grass from tropical central Africa that can survive several months of drought and grows even on saline and alkaline soils. It spreads by long, above-ground stems. It is a palatable, high quality forage when harvested or grazed regularly (every five weeks), with crude protein of about 8 percent. Star grass produces few seeds, so it must be propagated by sprigs from the main stem or from root rhizomes in well-prepared soil during rainy months. It can be grown successfully in a variety of environments: from sea level to altitudes of more than 915 meters (3,000 feet); in areas having a wide variation in rainfall; and in many soil types.

ERWIN KINSEY

▲ *Cynodon aethiopicus*

Star grass can be grazed or harvested for hay or silage, but it cannot withstand continuous heavy grazing. It may also be shaded out by taller grasses and trees. Larger types of star grass are suitable for cut-and-carry. It grows well with herbaceous legumes such as Centrosema, Desmodium, and Stylosanthes. Some varieties have a tendency to produce hydrocyanic acid.

Brachiaria

Brachiaria brizantha* (signalgrass), *Brachiaria mutica* (para grass), *Brachiaria ruziziensis* (ruzigrass), *many other Brachiarias
Brachiara species are common in most of Central and South America, sub-Saharan Africa and Southeast Asia. Brachiaria is a very productive perennial grass that grows in tufts with erect or semi-erect stems and with runners under the soil. Once planted, it spreads rapidly and can tolerate frequent grazing. It is used for permanent grazing, for ground cover against soil erosion and around water ponds. Bracharia species are usually relatively drought-resistant and remain green during a dry season of three to five months. They require moderate to fertile soils and do not tolerate waterlogging. While they have eight to 16 percent protein, they are not always palatable to goats.

PAUL RUDENBERG

▲ *Brachiaria*

Guatemala Grass–*Tripsacum andersonii*
(Honduras grass)
Guatemala grass is a tall, sturdy, broad-leafed perennial grass used very widely. It grows best in humid areas, but can be used in dry periods as it remains green during short dry seasons. It is used for cut-and-carry or hay, but is not suited for grazing. Guatemala grass is more persistent, but not as nutritious as Napier (elephant grass.) It can be used as a short living fence or for contour strips on hillsides, together with legumes. It grows best in fertile, well- drained soils.

ERWIN KINSEY

▲ *Tripsacum andersonii*

Guinea Grass–*Panicum maximum*
(Buffalo grass)
Guinea grass was originally native to Africa, but has been introduced to almost all tropical countries as a source of animal fodder. It is an ideal forage plant, growing well from sea level up to 2000 meters (6600 feet). Guinea grass also grows on a wide variety of fertile soils and can thrive under conditions of up to 30 percent shade, making it ideal for growing with other crops. Guinea grass requires a minimum of 900 mm per year of rainfall, but can survive dry spells and quick-moving fires. It also responds quickly to fertilizer and watering, but does not tolerate heavy grazing well.

ERWIN KINSEY

▲ *Panicum maximum*

Its fine, soft and very palatable leaves contain good levels of protein (13 to 21 percent). It grows up to 2 meters (6.5 feet) tall and bears fruit that are tiny rice-like seeds.

Napier grass–*Pennistum purpureum*
(Elephant grass, Uganda grass)
Napier grass is a fast-growing fodder grass that is easily propagated and produces large amounts of high quality forage. Best suited to high rainfall areas, Napier grass can also grow well in drier areas due to its deep roots. It does not, however, grow well in waterlogged areas. It can be grown with fodder trees along field boundaries, contours or terraces to help control erosion. It can also be intercropped with legumes and fodder trees or as a pure stand. Because of its fast-growing, aggressive nature and its ability to spread underground, it can become a weed in nearby gardens.

PAUL RUDENBERG

▲ *Pennistum purpureum*

Napier grass has a soft stem that is easy to cut for hay or for cut-and-carry systems. It should be cut between 15 and 26 cm (6 and 10 inches). Its protein content is reported at 24 percent. Used for hay, silage and grazing, Napier grass has tender, young leaves and stems that are very palatable for livestock. Older, more mature stems are generally rejected by animals.

Other important forage grasses include Timothy, crab grass, Pangola (Digitaria) and the panic grasses.

LEGUMES, INCLUDING NITROGEN-FIXING TREES

Acacia–*Acacia tortilis, Acacia mellifera, Acacia albida*
(Mimosa, Umbrella thorn, Senegalia)
Native to much of the Americas and also sub-Saharan Africa, the acacia tree is especially valuable as a dry season forage. It is very hardy and can grow on poor soils. In these conditions it may resemble a shrub. The acacia tree may reach 20 meters (65 feet) in height with an umbrella-shaped, flat-top canopy in maturity. Flowers range from white to a pale yellow.

ERWIN KINSEY

▲ *Acacia tortilis*

Goats browse the leaves of young trees. Although some species have thorns that are 2 to 10 cm (0.8 to 4 inches) long, goats are able to use their mobile upper lip to sort young leaves from the thorns.

Acacia leaves contain about 17 percent crude protein. High in fiber and protein, the pods of this species are also an important goat feed. Typically, goatherds shake the Acacia tree's limbs to remove numerous pods and feed their stock. Some trees yield up to 220 lb of pods (100 kg) in one season. Acacia trees also increase the fertility of underlying crops.

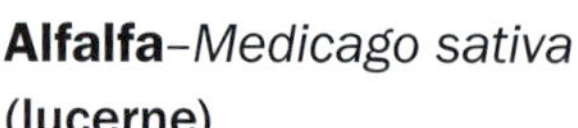

Alfalfa–*Medicago sativa*
(lucerne)
One the world's oldest crops, alfalfa is among the most prized of all pastures. It is grown worldwide as an excellent, highly palatable feed for many types of livestock. Alfalfa is high in protein (19 percent crude protein) and calcium, making it an excellent feed for milking animals. As a perennial crop, alfalfa can often persist for up to five years, and in some areas of the world may remain productive for much longer. It has both excellent frost and drought tolerance, surviving in regions with only 600 mm (24 inches) annual rainfall.

H.M. SHELTON

▲ *Menticago sativa*

Alfalfa is well known for its ability to improve soil structure. However soil conditions are very important in the successful establishment of alfalfa. It requires deep, fertile, well drained, alkaline to moderately acidic soil. The palatability of alfalfa is lower in the rainy season, but readily eaten in the dry season. Feeding large amounts of fresh alfalfa can cause bloat. This can be prevented by combining grass species with alfalfa in pastures.

Calliandra–*Calliandra clothyrus*
Calliandra is a small, thornless leguminous tree native to Central America and Mexico. It is rarely utilized in this region but it has been introduced to many tropical regions where it is used in agroforestry systems for fuel wood, plantation shade, as an intercrop hedgerow and more recently as livestock forage. It appears to be readily eaten by animals although only limited information is available on its nutritional value. Routinely fed to

ERWIN KINSEY

▲ *Calliandra clothyrus*

goats in Indonesia and Australia, it is reported to have 22 percent crude protein and 30 to 70 percent fiber in dried calliandra leaves.

Desmodium

Desmodium triflorum (**Amor du campo**); *Desmodium intortum* (**Greenleaf desmodium, Pega-pega, Kuru vine**); *Desmodium uncinatum* (**Silverleaf desmodium**)

Often called the "alfalfas of the tropics," the highly palatable and nutritious desmodium has many useful species. Some are trailing and climbing herbaceous varieties, and others are shrubs. While tolerance to drought varies, some grow in areas with only 700mm of annual rainfall. Desmodium is well adapted to tropical/subtropical environments and to a variety of soil types. It is tolerant of heavy grazing. And with 14 to 18 percent crude protein, it can markedly improve the diet of livestock on heavily grazed grass pastures.

ERWIN KINSEY

▲ *Desmodium triflorum*

Desmodium triflorum, pictured here, is low-lying, creeping forage with three-part leaves and pink blue-purple flowers. It can provide good ground cover for other crops during the wet season especially, if mown or closely cut.

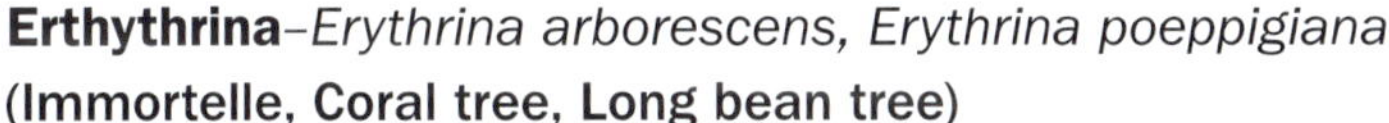

Erthythrina–*Erythrina arborescens, Erythrina poeppigiana* (**Immortelle, Coral tree, Long bean tree**)

Erythrina is a group of beautiful, large trees with orange or red flowers. Used often for shade, ornamentals or as living fence posts, most species are thorny. In humid tropical environments they are evergreen but become fully or partially deciduous in seasonally drier environments. Certain varieties are highly palatable and nutritious, and are used as a ruminant feed. They are well accepted by pen-fed goats, even as the sole feed. The leaves contain 26 to 30 percent crude protein.

PAUL RUDENBERG

▲ *Erthythrina*

Gliricidia–*Glicricidia sepium, Gliricidia maculata*

Gliricidia is a fast-growing, tropical, leguminous tree, which grows to between 10-15 meters (32 and 50 feet). Thought to have originated in Central America, it is one of the best-known multipurpose trees in that region. Now found in West Africa, the West Indies, Southern Asia and tropical North and South America, the tree grows best in warm, wet conditions but will survive with annual rainfall ranging from 800 to 2300 mm. Though gliricidia grows best on fertile soils, it will also grow well on acidic or high clay content soils. It can be established from cuttings or seed, (from seed leads to deeper roots). Gliricidia may be harvested at three month intervals to maximize foliage yield. Rich in protein (23 percent crude protein) and calcium (1.2 percent), it makes an excellent feed for goats for the dry season, since it contains sufficient levels of most minerals (except phosphorus and copper) to meet tropical livestock requirements.

PAUL RUDENBERG

▲ *Gliricidia in Haiti*

Leucaena–*Leucaena leucocephala, Leucaena diversifolia* **(Ipil Ipil, White popinac)**
Leucaena is a low to medium height 5 to 18 meters (16 to 60 feet), fast growing leguminous shrub or tree that can produce nutritive forage for goats and other ruminants throughout the tropics. It can produce twice as much edible dry matter as alfalfa, and be as productive as a forage tree for more than 20 years. Leucaena grows in both the tall type (leucocephala) and a short shrubby type (diversifolia). It will grow in areas with a wide variation of rainfall.

ERWIN KINSEY

▲ *Leucaena diversifolia*

It can easily be planted in hedgerows or alleys and on contours on hillsides, preventing soil loss while providing large quantities of high protein forage for cut-and-carry. In this system it is important to cut the leucaena regularly to provide palatable forage and to prevent it from producing seed. Leucaena has multiple uses and is suitable for firewood, timber, charcoal, fence posts, green manure or organic fertilizer. However reconverting to arable land is difficult as the roots are persistent.

KAI BAUER

▲ *Luecaena leucocephala used for forage and soil conservation*

WARNING

Leucaena is discouraged as a feed for pregnant animals as it can cause abortion. It contains an alkaloid called mimosine that can also cause a toxicity problem. Therefore it is best to feed it as no more than one-third of the goat's daily diet.

Stylo

Stylosanthes guianensis **(Brazilian lucerne, Graham stylo, Tropical lucerne)**; *Stylosanthes fruticosa* **(African stylo)**; *Stylosanthes humilis* **(Townsville stylo, Wild lucerne)**
There are many useful species of Stylosanthes, a forage found commonly in Latin America, Southeast Asia, Australia and Africa. Usually perennial shrubs with trifoliate leaves and branched stems, they are often hard to establish, but then have very good drought tolerance. Stylosanthes fruticosa (African Stylo), for example has hard seeds that remain in the soil for several years, allowing pastures to regrow annually. It has a low tolerance, however, to heavy grazing. The palatability and nutritive value (crude protein levels of 8 percent) are not as high as other more leafy legumes. Stylosanthes may have up to 6,000 kg (13,300 lb) per hectare dry matter production. It can be used to improve soil fertility and control erosion.

ERWIN KINSEY

▲ *Stylo*

Other legumes

Other important leguminous forages to consider in feeding goats include clover, vetch, birdsfoot trefoil, siratro, perennial soybean, *Vigna unguiculata* (Cowpea) and tropical kudzu, as well as the tree legumes: locust, *Sesbania grandifloria*, prosopis, *Cratylia argentea* and morenga (horseradish tree).

OTHER SELECTED FORAGES

Cassava-*Manihot esculenta*
(Manioc, Yucca)

Cassava is a very important tropical shrub, used extensively by hundreds of millions of people in many parts of the world. It has an average height of 1 to 3 meters (3.3 to 9.8 feet), with a palmate leaf formation. It is found in both “sweet” and “bitter” varieties, both of which are edible when properly prepared. Cassava produces bulky storage roots consisting of about 80 percent carbohydrates. People eat the “sweet cassava” tuber and process the bitter form to make flour and bread, among other foods. The stem, which is often burned for fuel, is also the planting material from which the roots and shoots grow.

ERWIN KINSEY

▲ Cassava

The shoots grow into leaves that are rich in proteins, vitamins and minerals. These leaves can be dried for livestock feed. Cassava hay is best harvested at an early growth stage, (three to four months), when approximately 30 to 45 cm (12 to 18 inches) above ground. Cassava hay has high protein content, 20 to 27 percent crude protein, and also 1.5 to 4 percent tannins.

ERWIN KINSEY

▲ Mulberry

Mulberry-*Morus alba*

Mulberry has considerable potential as goat feed, both from the biological and economic points of view. It has a high protein and energy content, similar to alfalfa. Mulberry also has good green biomass production throughout the year, including the dry season.

Banana-*Musa spp.*

Banana leaves and fruit are used as feed for goats by farmers in Tanzania, Uganda and other African countries. Young banana leaves have about 16 percent crude protein.

ERWIN KINSEY

▲ Banana

WARNING

For bitter cassava varieties, all parts of the plant contain a highly toxic poison (hydrocyanic or prussic acid). Animals should not be fed large quantities of fresh bitter cassava leaves. Sweet varieties have a lower concentration of the toxin. It is important to allow plant leaves to dry for two days before using as feed.

Goats Changed Vehbi's Family Living

▲ Emir with his Bukuroshja (Pretty)

Vehbi Hyseni is from Gadime village, municipality of Lipjan. He made application to receive goats from the multiethnic association, GRANITI. He needed milk for his children.

Gadime village is quite large and the members of the association GRANITI did not know this family. Then, the association members, together with the staff of Heifer Kosova, decided to visit this family. They listened to Vehbi's story about the war in 1999 and how he was deported and his 10 local goats were lost. Since that time his family has been suffering. Vehbi is poor and lives on social assistance with his wife and four children, two boys and two girls. One son, 10 year old Emir, is handicapped and he is in a wheelchair. The visitors noticed that Emir seemed very pale and weak.

The association members were satisfied that Vehbi had adequate forage to start a new goat enterprise. Vehbi took the Cornerstones Training and began to clean his barn and get ready for the goats. In September 2005, he received four Alpine goats Soon the family had fresh milk and cheese and yoghurt to eat and when the goats gave birth there were several male kids to slaughter for meat.

Vehbi says that since the first day they received the goats, Emir wanted to stay with the goats and asked his parents to be taken to the barn to see and help feed the goats. When the goats gave birth, Emir named one of the kids Bukuroshja (Pretty).

In their family, work is shared among the males and females. The males take care of the land (heavy work) whereas the females take care of the house and milking the goats and the children take the goats to pasture. Everyone in the family feeds the goats and takes care of the barn.

The goats provide three liters of milk a day for the family. And they help another family in the village by giving them milk. Emir is the one who seems to benefit most from the goat's milk. He moves about more freely, his face is not pale any more and he is often smiling. ◆

THE LEARNING GUIDE

REPRODUCTION

LEARNING OBJECTIVES

By the end of the session, participants will be able to:

- Select appropriate bucks and does for breeding
- Recognize the signs of estrus
- Practice natural breeding of goats
- Read a gestation table
- Castrate a buck kid

TERMS TO KNOW

- Reproduction
- Estrus/Heat
- Pregnancy/Gestation
- Flushing
- Artificial Insemination
- Hybrid Vigor
- Infertility
- Castration
- Selection
- Egg
- Sperm
- Fetus
- Rut
- Ejaculation
- Ovulation
- Open
- Anestrous

MATERIALS

- Mature does in heat
- Breeding bucks
- Cloths to make a buck rag
- Closed container for buck rag
- Young bucks that have been selected for castration
- Burdizzo castration tool or sharp knife
- 7 percent tincture of iodine solution

ADVANCE ASSIGNMENT

This lesson is best taught on a farm. This activity needs to be planned in advance in order to have does that are in estrus "heat" and breeding bucks. In some cultures separate lessons for men and women may be necessary since there is often sensitivity about mixed-gender discussions. The trainer may need to be the same sex as the participants. Use this opportunity to reinforce the similarities to human reproduction (i.e. ovulation, egg, sperm, cycles, pregnancy, giving birth). If castration is to be done, selected male kids should be available.

TIME (May vary according to group)	ACTIVITIES
15 minutes	**Group Sharing** Share important things that have happened during the past week.
20 minutes	**Get Everyone Thinking and Talking** ■ Ask farmers to share their experiences in breeding goats – what works and what does not work well. Talk about what happens when a doe goes through a season without being bred.
90 minutes	**Observation of Bucks and Does (Estrus, Breeding, Etc.)** ■ Select bucks and does. ■ Discuss seasons of the year when goats commonly breed in this area and why. ■ Review signs of estrus and observe does that may be in heat. ■ Review breeding behavior. ■ Bring selected buck to the breeding area. ■ Prepare a buck rag for later use. ■ Bring the doe to the buck to breed (in a fenced area). ■ Discuss gestation table and estimate date for kidding.
60 minutes	**Practice Castration** ■ Gather materials (Burdizzo clamps or a flat hard metal piece or scalpel blade can be used in the same way. Castration by cutting can also be practiced. Be sure to have 7 percent tincture of iodine available.) ■ Review common, relatively safe castration methods and where to obtain needed equipment. ■ Review the dangers of tetanus and screwworm – both possible with all methods (including Burdizzo) – and how to avoid them.
2 hours	**Advanced Activity – Practice Artificial Insemination (AI)** ■ This activity needs to be planned well in advance. ■ Flushing does in advance may help bring them into heat. ■ Choose does that are to be bred and have them next to the buck's pen for several days. ■ An AI technician should be present to conduct the exercise. The AI technician should provide the necessary equipment for the training session. ■ This activity should be conducted in the cool of the day.

REVIEW-20 MINUTES

- What was useful?
- What was a surprise?
- What did not work out well?
- What do you know now that you did not know before?
- What can you do as soon as you get home?
- What practices will be difficult to do at home?

THE LESSON

REPRODUCTION

Reproduction is the biological process by which a new individual of the same species is created. For goats, this process begins with mating of a buck and a doe or artificial insemination of a doe. When successful, the result is conception, whereby an egg from the doe's ovary is fertilized (or joined) by a sperm from the buck to create a new individual. Conception is followed by a pregnancy or gestation period of about 150 days during which the fetus (the unborn kid) grows inside the doe's uterus until birth.

In temperate climates, goats are seasonal breeders, with most mating in the period between late summer and early winter. During this period, the doe comes into heat, or estrus, every 18 to 22 days (average 21 days). Estrus or heat in the doe can last for as little as a few hours or as many as two to three days. Certain does may cycle (come into estrus) at other times during the year, particularly if a new buck is introduced to the farm, or if artificial lighting is used. In tropical climates, female goats usually cycle throughout the year.

With intensive management, a doe can be bred three times in a two-year period. After the five-month gestation period the doe will "kid." Twins are common, but single or triplet births are also typical. Where nutrition is poor and breeding is random, singles are the most common. After kidding, does are usually milked for 10 months. Once bred again, a lactating doe should be "dried off" (milking should be suspended) two months before the kidding date.

BREEDING AND SELECTION

A successful goat enterprise depends on regular breeding and healthy births to expand the herd. Breeding must be effectively managed, as uncontrolled mating can limit productivity. Managed breeding involves selecting suitable dams and sires (mothers and fathers), and choosing an appropriate time for mating. It also includes managing nutrition during the breeding period to increase conception, and scheduling breeding so that does give birth when feed is abundant. In tropical climates, the best time for kids to be born may be at the end of the rainy season when forage is abundant. In temperate climates late winter or early spring may be best.

Improving a goat herd is realized through effective animal management and genetic selection using established criteria. Management means ensuring animals have good general care, balanced nutrition and preventive health care. Genetic selection involves, among other things, carefully choosing the physical traits of the animals to be bred as a means of generating desirable qualities in their offspring. The combination of these two (management and selection) will contribute to a strong, healthy and productive herd.

There are several types of breeding. These are a listed below along with their main purpose and advantages. In choosing one over another, carefully consider the final objectives for the offspring: Is the aim to build to a healthy dairy or meat herd? Is the aim to sell offspring?

Pure-breeding

Pure-breeding is the mating of two purebred animals of the same breed. The offspring of two purebred animals will possess the characteristics of that breed, while possibly being deficient in other characteristics, like disease resistance. The offspring will also be purebreds, and may have a higher sale value. However, using a purebred buck as herd sire with purebred does is not always a guarantee of success. In addition to records on the buck, it is important to have the production records of his dam or daughters.

Cross-breeding

Cross-breeding is the mating of two goats of different breeds, or breeding an exotic goat with a local goat. First generation cross-bred progeny are often superior to the average of their parents for production traits such as milk yield and growth rate. They may also be hardier than their parents. This improved performance is known as hybrid vigor. Cross-breeding is often used to improve the productivity of local animals. Using an exotic (improver) buck to breed a local doe will produce strong, vigorous offspring. These offspring will have increased production and also some resistance to local diseases. If you cross a dairy and meat breed, the female offspring will produce more milk than the pure meat breed, and wean heavier kids than the pure dairy breed.

Line-breeding

Line-breeding is the mating of related individuals (not closely related) in order to keep a common ancestor in the pedigree. This increases the chances of maintaining a desirable trait in the offspring.

In-breeding

In-breeding is the mating of animals that are closely related. The more closely related two parents are, the greater the risk that their offspring will carry or replicate a genetic defect or weakness present in their family ancestry. Offspring that inherit their genes from two closely related parents tend to be less productive and less hardy. This is called "in-breeding depression." A breeding program should always include a check of well-kept records to prevent breeding goats to their parents, full siblings, half siblings or first cousins.

WARNING

Do not breed two polled (hornless at birth) goats. Breeding two hornless animals may result in genital abnormalities in the offspring.

DEVELOPING A MANAGED BREEDING PROGRAM

Select Foundation Stock

When developing a managed breeding program, begin by selecting a breed for improved goat production, and foundation stock from this breed, based on four primary factors: adaptability to environment, reproductive rate, growth rate and milk production or carcass value (depending on your herd objective).

Identification and Record-keeping

Identify each goat in the herd and record their sire and dam (father and mother). Be sure to use a unique name or number for each individual in order to have accurate information on the animal's lineage. These controls are useful in preventing unwanted line- or in-breeding that could cause birth defects.

Choose Relevant Traits

Examine the herd and make a short list of the heritable (transmittable) traits, both desirable and undesirable, that are found in the goat herd. These traits will be important in selecting the herd's dam and sire. Milk production and ability to forage, for example, are two heritable traits that are important.

Which traits found in the herd are most important to preserve?

- Choose does that are easy to breed successfully and that are good mothers.
- For milk, choose only bucks with sound conformation that are large for their age and are from a doe that was one of the top milk producers in the herd.
- For meat, choose large capacity animals with good muscling.

Which traits need improvement in the herd?

- If goats in the herd limp or have weak legs, include good leg conformation as one of the selection traits.
- If does in the herd get mastitis because the conformation of their udders is big and droopy, include good udder conformation as one of the traits.

By making strategic choices of traits on this list, the farmer can, through the breeding and culling program, either propagate desired traits, or breed out undesired characteristics.

Establish Selection Criteria

Based on the chosen traits from step three listed above, establish selection criteria for herd foundation stock, breeding stock, production (dairy or meat), and for culling and castration. This will help to determine how each animal in the herd can be best used. Criteria will range from growth rates (for kids), resistance to certain diseases, to adult size and average milk production. Specific criteria should allow for differences due to age and litter size. These criteria are then implemented in the herd.

For breeding stock, for example, the selection criteria might be that kids are healthy, fast-growing twins from the top milkers or meat producers in the herd. This selection can be made as early as the time of weaning. Keep in mind that twins grow slower than single kids; so, twins should be compared to twins and singles to singles when determining the fastest growers.

For dairy production, the criteria will likely be high milk production, no cases of mastitis, and good udder conformation. Remember that the volume each doe produces varies by age; younger does produce less milk than 3- or 4–year olds.

When choosing which dairy does or bucks to cull, the criteria might be poor milkers, crippled goats and does that fail to breed or are not good mothers. Bucks should be housed separately from their offspring to avoid in-breeding. They may be moved to another farm or traded for another buck.

Develop Local Breeding Resources or Utilize Existing Support
Form a "buck circle" with other nearby goat farmers or communities. Nearby farms can start out rotating a set of three or four bucks through their communities. Each community or farm should plan on keeping one or two buck kids that are offspring of these breeding bucks and from the best does. These buck kids can then be used on unrelated does or transported to more distant communities for use.

Often a government farm or a larger farm is used as a genetic center. Periodically, daughters from the most superior does in the villages are purchased and brought to the center to be bred to bucks that have also been identified as superior. Offspring of these bucks are evaluated. The very best bucks are then retained at the center for a few years before going out to the villages. The least productive bucks are culled. High quality bucks, sons of the very best bucks, and in some cases, daughters, go out to the communities.

THE BREEDING BUCK

The breeding buck is the male goat chosen to father the kids in a herd. Although bucks may be chosen within the farm's own herd, many small herd owners decide not to keep their own buck. Often the owner of a breeding buck will allow his animal to mate with does or another farm in exchange for a breeding fee. Only bucks from high quality parents should be kept for breeding purposes.

As a general rule, breeding bucks should have a large body capacity to provide ample digestive capacity, strength and vigor. Good overall conformation is important, as well as normal and symmetrical reproductive organs.

In selecting a breeding buck for a dairy herd, use a buck that has already produced high quality milking daughters. For the first year of breeding, judge a buck by his mother's milk production. Ideally, the buck's dam should give as much or more milk than the doe's mother. A dairy breeding buck should have dairy character and show animation, good body capacity and general openness (ribs that are long and far apart). A buck selected for meat production should show strong legs and good muscling through the shoulder, hindquarter, back, loin, rump and inside the rear leg. If available, request records of the buck's average daily weight gain in order to estimate the growth rate of his offspring.

REPRODUCTIVE ORGANS OF A BUCK

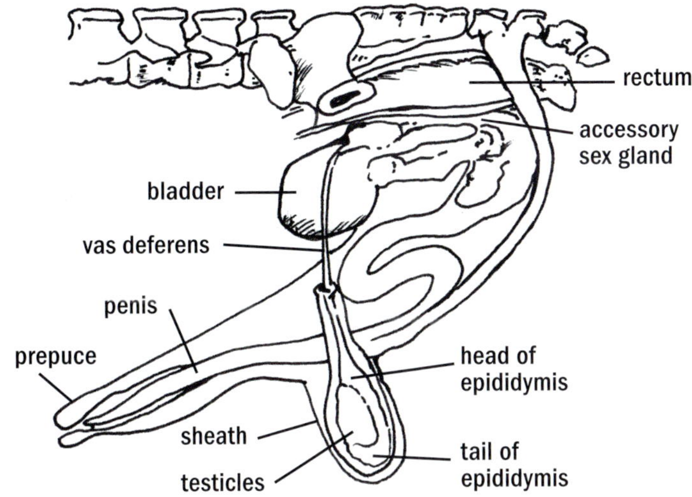

If breeding for pure-bred kids, select a buck of the same breed as the doe. When cross-breeding, it is best to breed a local doe to an exotic (improver) buck. For increased vigor breed two goats of different breeds, i.e., Nubian and Toggenburg. However, the offspring cannot be sold as pure-breds.

Good nutrition is essential before and during the breeding season. A 60 kg buck needs 0.5 kg (1.1 lb) of concentrate per day. Plenty of water and green chop (forage) should be offered free-choice. If concentrate or legume hay is not available, be sure the buck has adequate forage and is allowed to browse extensively.

During breeding season, bucks mature and go into rut or breeding behavior, making them more prone to fight each other. Separating bucks during this period is advisable. In this season, bucks will also exude a strong musk-like odor that can be used to stimulate estrus in does. Because of this odor, bucks should be kept away from milking areas and areas where milk products are processed.

Other characteristics of breeding behavior include repulsive habits, such as urinating on their front legs and/or face. This indicates readiness for breeding, and generally the habit stops when the doe is no longer in heat.

THE BREEDING DOE

A doe should be over ¾ of her mature weight at breeding, usually seven to 10 months old and in good physical condition, but should not be fat. The heat (estrous)* period usually lasts from 48 to 72 hours and ovulation occurs 24 to 36 hours after the onset of the heat period. The doe should be bred during the last half of the estrus.* Breed on the second day and repeat in 12 hours if she is still in heat, as this increases the chances

REPRODUCTIVE ORGANS OF A DOE

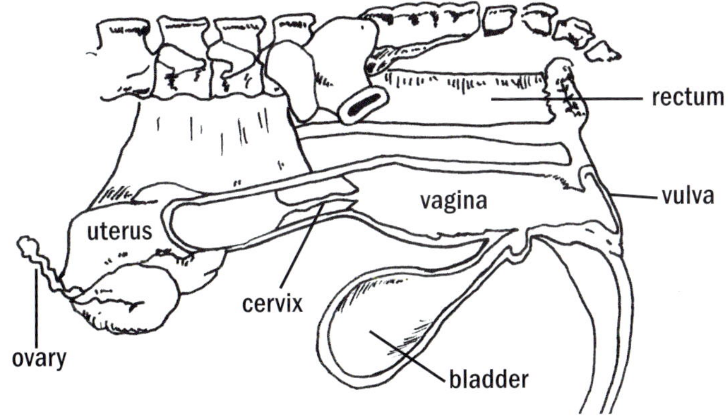

*Note: Estrous is an adjective. Estrus is a noun.

Signs of Estrus (Heat)

Failure to properly recognize and detect signs of estrus is a major cause of poor reproduction. To be successful, a farmer must know the signs of estrus and breed the does in a timely manner.

The main signs of estrus are detailed below:

- Swelling and redness of the vulva
- Mucous discharge from the vulva
- Flagging of the tail
- Nervousness and bleating
- Mounting and being mounted by other does
- Frequent urination
- Willingness to stand for the buck

WARNING

In the tropics or hot weather, breed early or late in the cool of the day to preserve the quality of the sperm. Remember the egg is released during the final period of estrus, so breeding should take place then.

Encouraging a Doe to Come into Heat

The introduction of a buck or his scent at the beginning of the season will often bring an entire group of does into heat in about eight days. To do so, place the doe near the buck or rub a piece of cloth on the buck's head behind the horn area where musk glands (odor producing glands) are located. Bring the cloth to the doe and let her smell it for several minutes once a day. After using the cloth, put it in a sealed jar or plastic bag to keep the scent.

Although a buck or its scent is an effective method to induce heat, most does will cycle without the buck's presence. The season can also be induced and optimized by decreasing light and temperature. Shorter days encourage heat cycles. Increasing the plane of nutrition at the time of breeding, called "flushing," is also beneficial. This is done by feeding high quality, lush pasture or additional concentrate, or protein supplements. This may also increase the number of ovulations and the likelihood of multiple births.

Natural Breeding

When the doe is in heat, she should be taken to the buck. If she is in standing heat, she will let him mount and breed. Males are most fertile in the first three to four breedings of the day. After that a buck's semen will have fewer sperm, decreasing the chances of successful breeding. Bucks should not be left with does for prolonged periods. The farmer should observe the breeding and record both the doe and buck's identification numbers and the date and time of breeding.

Breeding Behavior

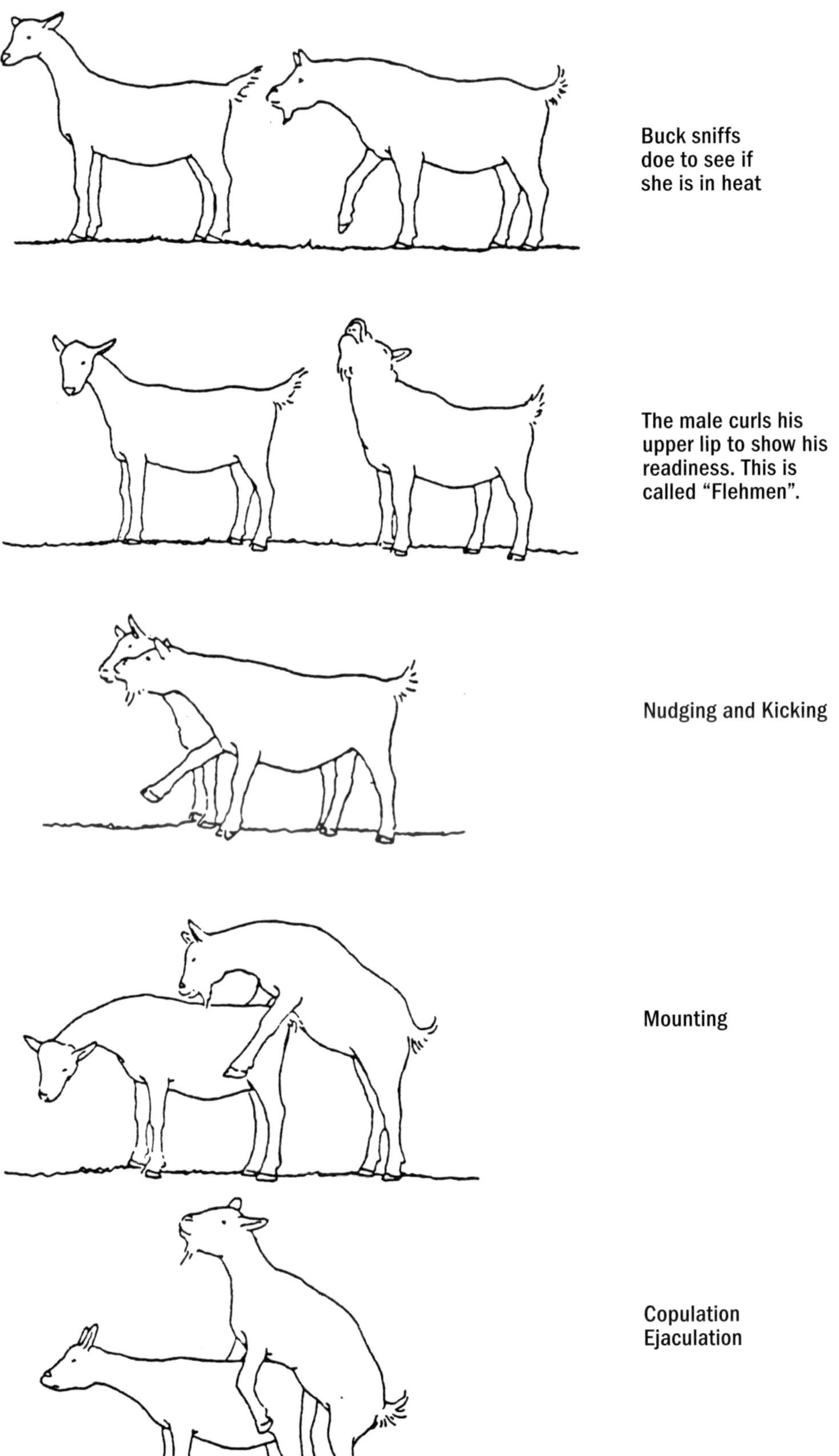

Artificial Insemination

Artificial insemination (AI) in goats is a well-developed technology with a history of more than 100 years. It involves collection of semen from a buck and its transfer to the reproductive tract of the doe. Either fresh or commercially available frozen semen can be used to improve the genetics of offspring. Semen can be collected from top sires, frozen, and then transported throughout the world, where it can improve large goat populations. Collection, freezing and storage of semen must be done by a trained technician.

With good heat detection, records and semen handling, AI will greatly help to increase the rate of genetic improvement. However, it may decrease production unless superior management, nutrition, housing and parasite control are already in place. There must be good access roads to the village and ability to communicate by mobile phones to reach the AI technician in a timely fashion.

ARTIFICIAL BREEDING (AI)	NATURAL BREEDING
Increases rate of genetic improvement by use of superior buck semen	Genetic improvement is less rapid
Eliminates costs of feeding, housing, separate fencing and labor for rearing one or several bucks	You must feed and house the buck all year or pay a breeding fee for using another farmer's buck
Numerous does can be bred in a single day	Three to four does can be bred in a single day
Avoids most reproductive diseases	Bucks may spread infectious disease
It is easier to know pedigrees	Higher pregnancy rate
Requires excellent skills of heat detection and technical infrastructure for semen collection, storage and insemination	Does not require technical knowledge of artificial insemination

General Consideration for AI in Goats

Obtain frozen semen from a reputable source. If using fresh semen, cervical insemination during natural heat gives better results. If using frozen semen, synchronization of heats is more effective. One of the simplest means of synchronization is the sudden introduction of a buck or his odor early in the fall.

Hormone injections like Pregnant Mare Serum Gonadotrophin (PMSG) and Follicle Stimulating Hormone (FSH) can bring does into heat. These are powerful medicines which can only be used by a veterinarian or trained technician. The dose of PMSG used depends on the age of the goat, the current milk production and the season of the year. Progestin vaginal sponges can also be used in certain countries.

WARNING

Pregnant women should not be in the presence of these reproductive hormones.

Optimum Insemination Time

Accurate detection of the start of estrus is important while implementing an artificial insemination program. This is because AI is most effective just prior to ovulation, which occurs near mid or late estrus. A doe can be inseminated at approximately 12 hours after the first sign of estrus and this can be repeated two times at 12 hour intervals to increase chances of pregnancy and of multiple births. If the does are in estrus (heat) in the morning, inseminate in the evening. If estrus (heat) occurs in the evening, inseminate the following morning. Repeating AI can be costly. You may want to use a "clean-up buck" to breed a doe if she does not conceive and returns to heat after AI.

Materials and Equipment for Insemination in Goats

- Liquid nitrogen tank
- Speculum
- Lubricant
- AI gun
- Sheaths
- AI light
- Facilities for restraining
- Straw tweezers, straw cutter
- A thaw box
- Thermometer
- Paper towels, etc.

Thawing of Semen

Methods for semen thawing vary among AI equipment manufacturers, and it is best to follow their recommendation. If the recommendations are not available, remove the frozen straw from the liquid nitrogen tank with straw tweezers, and place it into a thaw box filled with warm water at 37° Celsius (98.6°F) for 30 to 60 seconds. This requires a source of warm water and an accurate thermometer. After thawing, dry the straw thoroughly with a paper towel. The semen must be kept warm and must not be exposed to sunlight or water during the inseminating process to prevent damaging or killing sperm cells. The thawed semen should be used within five minutes.

Preparation of Straw

- An inseminating gun, designed especially for the type of straw is needed. Have cover sheath available, sealed until needed.
- Pull plunger on gun back 4 to 6 inches and insert straw into gun, cotton plug end first (towards plunger).
- Hold gun in upright position, allowing air bubble to rise to the sealed end.
- Cut sealed end of straw with scissors. Take care to cut straw squarely for proper seating.
- Install the sheath over the gun, fastening it down with the provided O-ring. Install it so that the wider side of the ring faces the straw, with the narrower side facing the syringe end.

Insemination Procedure

- All the care in handling, storage and preparation of semen will be useless if the inseminating process is not done carefully and cleanly. Be sure to restrain the doe to be inseminated either in the breeding stand or other suitable facilities.

It may be necessary to lift the doe's hindquarters if she will not stand. Before proceeding clean the doe's vulva with a dry paper towel. Place light inside speculum. Turn light on and insert the sterile and properly lubricated speculum slowly through the vulva and into the vagina in gentle manner.

- Complete insertion of the speculum. Locate the cervix visually. The cervix should be reddish purple in color and covered with white mucus if the doe is truly in heat. Center the speculum over the opening of the cervix.
- Insert the inseminating gun into speculum to the opening of the cervix. Gently manipulate the insemination gun in circular motion through the cervical canal, penetrating about 1.5 inches (4cm) past the cervical opening.
- Deposit the semen near the uterine end of cervix or just inside the uterus. Deposit the semen slowly taking at least five seconds.
- Remove the instrument without releasing the syringe and then carefully remove the speculum.
- Record all important pertinent information in breeding record.
- Discard all disposable materials, then clean and sterilize reusable materials.

Be sure to check for signs of heat within 18 to 22 days to be sure your doe is bred. If she is not bred, repeat AI or use natural breeding.

GESTATION CHART

BRED	BIRTH*
July	December (-3)
August	January (-3)
September	February (-3)
October	March (-1)
November	April (-1)
December	May (-1)
January	June (-1)
February	July (-0)
March	August (-3)
April	September (-3)
May	October (-3)
June	November (-3)

* To determine day due to birth, take day bred and subtract the number listed after the month in the right column from the day she was bred. This will give you the day due in the month listed as the birth month. For example, if bred July 10th, the doe would be due to birth December 7th. If bred November 20th, she would be due to birth on April 19th.

CARING FOR THE PREGNANT DOE

Feed the doe concentrate containing protein and energy during pregnancy to keep her healthy and to insure strong kids. As the pregnancy progresses, gradually increase the nutrients in the doe's diet without letting her get fat. In the last three weeks prior to kidding, replace high calcium legumes with grass hay. This forces the animal to mobilize its own body calcium reserves and prepare for milking, and also helps prevent hypocalcaemia, or milk fever. For more information on milk fever, see the health lesson.

Give the doe Clostridium Perfringens C & D vaccine and Tetanus Toxoid three weeks prior to kidding. Repeat after 30 days if the doe has not been previously vaccinated. If recommended in the local area, also give a vitamin E and selenium injection during the dry season to prevent white-muscle disease or infertility. Selenium plays an important role in regulating female reproduction. Give the correct dose, as too much selenium is toxic.

Does frequently have high parasite populations at the time of parturition (kidding). Some breeders choose to deworm seven to 10 days before kidding while others deworm at the time of kidding. Having a fecal sample tested to determine parasite load can help in planning an overall parasite control program. (See Lesson 8 for parasite control.)

Breeding Back the Doe After Kidding

In temperate areas, does breed in the fall and give birth in the spring when feed is plentiful. Does will not return to heat until the next fall when daylight begins to decrease. As mentioned before, some breeders use artificial lights to make their does cycle even in the summer. This extra expense can pay off if the goat herd has excellent nutrition and health care. A doe can be bred again eight weeks after giving birth if she comes into heat.

In tropical countries where daylight does not change much throughout the year, does may come into heat at any time. Again with excellent nutrition and health care, does can become pregnant eight to 12 weeks after giving birth and can kid three times in two years. This is only possible with a reliable year-round feed supply. In the tropics dry periods may limit feed. If there are long dry periods with poor forage, keep the doe open (not pregnant) until there is a good chance of adequate feed when she kids. Otherwise the dam and kids could die from lack of nutrition.

INFERTILITY IN GOATS

The term infertility simply means a failure to get or stay pregnant. It can be caused by a wide range of factors, including specific nutritional deficiencies or general lack of nutrition, infectious diseases, genetic factors, stress, toxic plants and seasonal differences. Infertility is also related to a number of situations that cause reproductive failure such as failure of a farmer to detect estrus, lack of estrus (anestrus), hereditary genital defects, embryonic death and abortions.

Common causes of infertility and reproductive failure are:

- *Nutritional Deficiencies*
  Although goats are hardy animals that can thrive in very harsh environments, adequate nutrition is essential to reduce the prevalence of infertility, especially anestrous. In developing countries, general malnutrition is probably the major cause of infertility. More specific deficiencies of protein, energy, vitamins, and minerals like phosphorus, copper, iodine, and zinc can cause infertility. For this

reason adequate supplementation of all goats with mineralized salt or mineral blocks is important to reproductive health.

- ***Infections of Genital Tract***
  The genital tract of goats can become infected with bacteria, viruses, fungi and chlamydia agents. Some of these pathogens can cause infection resulting in death of the embryo and abortion in pregnant animals and may subsequently lead to infertility. Some of the common infectious agents causing abortion are chlamydia, brucella, toxoplasma, listeria, foot and mouth disease, mycoplasma and salmonella but there are many more.

- ***Failure to Detect Estrus***
  In many management situations, poor observation skills or lack of full knowledge of the signs of estrus by those caring for livestock day-to-day may lead to a failure to detect goats in heat. This can lead to a costly low reproductive rate in the herd.

  A doe that does not get pregnant or stay pregnant despite being mated several times to a fertile male is called a "repeat breeder." This condition may be from infections, hormonal imbalances, or hereditary and management factors. When evaluating a doe for infertility or as a repeat breeder, it is important to remember that the buck may also be infertile, either temporarily or permanently. If many does are mated to a breeding buck, but do not become pregnant, try another buck.

- ***Anestrus***
  Anestrus is when a female does not exhibit heat. This happens due to a change in the environment, heat stress, nutritional deficiencies, lactation stress or age. Low levels of dietary energy or protein, mineral deficiencies (phosphorus, manganese, cobalt, copper) and vitamin deficiencies (vitamin A and E) may suppress the signs of estrus or cause irregular estrus.

CASTRATION OF BUCKS

Castration is the process by which a goat's spermatic cord is severed to prevent breeding.

It is important to keep good records to determine which animals are productive and disease-free. You may want to sell your best bucks as breeding animals. You should castrate all bucks that you do not plan to use for breeding. These bucks can be used for meat. This will prevent undesired breeding, since bucks as young as three months of age may breed. You can castrate anytime from two weeks to two months; however, you will get better growth if you wait until nearer two months and use the closed method. This may also help reduce or prevent urolithiasis in the castrated bucks, as the development of urethra and urethra process is dependent on testosterone.

There are two main methods of castration. A tetanus anti-toxin shot is recommended prior to either method, or give penicillin if anti-toxin is not available.

Closed Method or Castration With the Burdizzo Clamp

This method is preferred because there is less risk of infection, fly strike (maggots), or tetanus. It is used most commonly in animals over six weeks of age. However the tool is expensive and sometimes not available.

To use the Burdizzo clamp, have someone hold the buck on his rump with the testicles laying between the rear legs. First, feel for the spermatic cord and blood vessels beneath the skin between the testicle and the abdomen (see diagram). Close the clamp slightly, then firmly over the cord, remaining on one side of the middle. Repeat in a slightly different position on the same side. Be careful not to clamp across the middle as you can crush the urethra! Repeat on the opposite side to crush the vessels on each side separately. The goat may be uncomfortable for a short while. After a few weeks the testicles will becomes smaller and firm.

Open Method

This method is used most often in kids that are 2 to 4 weeks of age.

Materials needed:

- sharp knife, razor blade or scissors that have been sterilized
- alcohol
- iodine
- soap and water
- person to help

Place the materials needed on a clean surface. Wash hands with soap and water. Wash scrotum with soapy water. Dip the scrotum in either alcohol or iodine (7 percent tincture of iodine).

Procedure:

- Make a 1 cm vertical cut on each side and push testicles out through the opening. You may need to cut the shiny white tunic, also.
- Gently pull the testicles until the cord is severed (tear it, do not cut).
- Dip scrotum in 7 percent tincture of iodine or other antiseptic.
- Release kid but observe periodically. Allow to nurse or bottle feed immediately, but keep away from other rowdy kids. In most situations the incision will heal naturally within a few days.

WARNING

Try to avoid fly season when you are doing these procedures and spray for flies as needed. Flies could cause a maggot infestation if not checked.

Thecla Makes a Difference

Thecla Makoke has always been among a minority of women in a field dominated by men. Having studied farm management in Tanzania and Holland in the mid-1980s, she often felt the challenge: why was she encroaching into the male territory of agricultural development studies. But she persevered. "I know that my mother and aunts always did most of the farm work, so why should their advisors be men?" she asked. Twenty years later she is an experienced and accomplished technician, well respected among her peers.

The women's groups in the area of Kibaha have gained her respect as well. In this region of southeastern Tanzania, coconut, cashew nut, citrus and mango trees are interspersed among the cassava, maize and bean fields common throughout the undulating terrain. Even in the driest years, there is shade under the tree foliage where people can do farm work at midday. In the fields and around homesteads, young children help their parents and grandparents. Only 50 percent of school-age children are enrolled in school. Some cultivate around the pineapples; others move their local chickens out to graze. Others wash clothes or household utensils and dry them in the sun on small bamboo stands.

Women's groups from five villages near Kibaha in southeastern Tanzania approached the district office of Heifer International, where Thecla worked, to obtain animals for producing milk for their children. Thecla reported: "They were surprised when we suggested dairy goats, since almost no one in the area had ever drunk milk from a goat. Through our encouragement, and a promise of assistance from Heifer International, BOTHAR in Ireland and Heifer Netherlands, the project began. Twenty five families in each village were encouraged to prepare goat sheds, plant fodder grasses and trees, and to attend training on husbandry. Twenty-five percent of these households are headed by women."

With Thecla's encouragement, the families began preparing by planting fodder and building small goat sheds. "Some family household heads did not believe that we could be trusted to fulfill on the promise of dairy goats," she said. "Most of the women were more trusting, but some of them were prevented by their husbands." After persistent visits and hands-on training, 113 families finished preparations, and each received a dairy goat between 2002 and 2003. Each family signed a contract to pass on three offspring.

One such family is that of Tikiti, a farmer and a single mother with two young girls in the village of Mbwawa. They make their living on a two-acre farm. Mbwawa (Reservoir), lies on a dirt feeder road extending north from Mlandizi town, about 40 miles inland along the main TanZam Highway which traverses Tanzania. The predominately Muslim population makes its livelihood from farming, but increasing numbers of rural youths have moved to the nearby urban centers of Dar-es-Salaam and Kibaha for work. When you visit Tikiti, you will find her busy selling palm wine along with small, spicy deep-fried ground pigeon pea cakes called bajias, an income-generating project she runs together with her neighbor.

Tikiti has done well with the dairy goat she received in 2002. Her goat, named 'Tegemeo' (Confidence), has produced two sets of twins, two males and two females. Tikiti has sold two of the offspring for $70 each, and can sell all the milk that they do not consume at home, for $US 0.40 a liter. This income has enabled her to join a small savings and credit group with other neighbors, from which they can obtain funds when needed.

She is not alone. The milk and sale of goat offspring are helping to raise the nutritional and income levels of participant Kibaha families. Already 23 offspring have been passed on to new families. The project that Thecla oversees is making an unimaginable difference in these families' lives. ◆

THE LEARNING GUIDE

KIDDING

LEARNING OBJECTIVES

By the end of the session, participants will be able to:

- Describe the signs of parturition (kidding)
- Recognize normal and abnormal birthing positions
- Prepare a kidding kit
- Give supportive care to a doe during kidding
- Disbud kids
- Develop a personal plan for caring for the doe and newborn kids

TERMS TO KNOW

- Parturition
- Colostrum
- Supportive care
- Umbilical cord
- Placenta
- Lochia
- Cervix
- Water sac
- Navel
- Uterus

MATERIALS

Posters

- Two poster-sized pieces of paper and tape
- Marking pens or dark pencils
- A local artist

Materials for birthing kit

- Bucket
- 7 percent tincture of iodine in small bottle
- Ear tags
- Soft cloths
- Sharp knife
- Tetanus/clostridium C&D vaccine and syringes
- Ear tags or tattoo kit
- Pen, pencil, paper for recording birth

Disbudding Kids

- A disbudding box (an example of what will be made)
- Young goats under two weeks of age for disbudding
- Disbudding iron
- Precut wood, nails and hammers to make disbudding boxes as per description in text
- Nursing bottles with milk for supportive care after disbudding

ADVANCE ASSIGNMENT

Based on conversations with the group, decide what the most urgent problems are related to pregnancy, kidding and raising newborn kids on farms in this community. For two of these common problems, you will need to have someone draw a pair of sketches, illustrating good management and poor management that lead to this problem. See examples at end of this Learning Guide. This is a long lesson and you may want to decide to conduct it in two or three different sessions.

TIME (May vary according to group)	ACTIVITIES
15 minutes	**Group Sharing** What has happened since the last session? Have you made any changes in your goat management practices? Today we are talking about kidding. What do you need to learn from this session? List responses.
45 minutes	**Get Everyone Thinking and Talking** Have someone show each of the drawings to all members of the group. They will need to slowly circulate in the group to be sure that everyone sees them up close. Then post the drawings at the front and begin the discussion. ■ What did you see in the first drawing? In the second one? ■ How are they different? ■ In the first picture, what do you think happened? ■ What would you change to make the outcome look more like the second picture? ■ Which of these pictures looks more like a farm in this community? Why? ■ Make a list of practices that would make the first picture look more like the second. This is the time to present appropriate new concepts from Lesson 6, and your experience related to the subject of kidding and care of the newborn. Present a few relevant terms from the "terms to know" list and ask participants to describe them in their own words.
60 minutes	**Normal and Abnormal Birth** ■ Look at the illustrations in the training guide and discuss. Observe a doe kidding if possible. Be ready to sex the kid, dip the kid's navel and get the kid nursing. You may also want to ear tag or tattoo the kid and fill in the record sheet. Let the group decide on the role of the farmer during kidding. Talk about supportive care. If you are meeting on a farm, you can stop any session at the point a doe will be giving birth and discuss afterwards. ■ Warning: Women of childbearing age should always wear gloves when assisting births, touching amniotic fluid or placenta. Otherwise they can get toxoplasmosis or leptospirosis which can severely damage or abort any fetus they are carrying.
20 minutes	**Assemble a Kidding Kit** Discuss why each item is needed in the kit.
20 minutes or 1 hour if building box	**Build a Disbudding Box and Disbud Kids** If time is limited the class members may be given the materials to build the box at home. The disbudding exercise can be done with a disbudding box that is already finished.

30 minutes	**Homework** Ask farmers to prepare their own plan for caring for their does and newborn kids.

REVIEW-20 MINUTES

- What was useful?
- What was a surprise?
- What did not work out well?
- What do you know now that you did not know before?
- What can you do as soon as you get home?
- What practices will be difficult to do at home?

EXAMPLES OF ILLUSTRATIONS

LOSS OF VERY YOUNG KIDS TO DISEASE

Make two drawings of newborn goats with their mothers.

- *In one drawing, show one skinny kid nursing from a baby bottle, with the doe (with placenta tag) standing off to the side a little. Show the doe sniffing a second kid which obviously appears sick. The pen should have a dirt floor with some manure and water on it, and no straw. Show a water bucket on the floor of the pen.*

- *In a second drawing, show a healthy newborn kid happily nursing the mother. The pen floor should have straw. Show the water bucket on a hook. Show the farmer holding the second kid with the iodine jar turned against the navel. Place some feed in a feeder near the doe.*

Or make drawings of example two or another example that is important to the group.

DYSTOCIA (DIFFICULT BIRTH) LEADING TO LOSS OF KIDS AT BIRTH.

- *In one drawing show a small, thin doe with her head hung low. Show a kid that has died lying on the floor in a pool of fluid. Show a worn out, sweating farmer in the pen holding a second kid. The pen should have a dirt floor, some manure, water and no straw.*

- *In the second drawing, show a normal birth with a healthy, nourished mother licking a beautiful newborn kid. The pen floor should be clean and covered with straw.*

THE LESSON

KIDDING

INTRODUCTION

A healthy birth requires planning and preparation, balanced nutrition and preventive healthcare during pregnancy, and supportive care at the time of kidding and in the following days. To prepare for kidding, it is important to know when the doe will kid. Establishing the date is important because it allows the farmer to plan precisely when to:

- Prepare a birthing kit
- Set up a kidding pen (for intensive management systems)
- Bring does from pastures or mountains to near the farm house
- Adapt a doe's diet to prevent milk fever and pregnancy toxemia
- Watch for signs of parturition

Utilize the Gestation Table to estimate the doe's kidding date (see Chapter 5).

BIRTHING KIT

Regardless of the goat management system in place, a birthing kit is excellent preparation for healthy or abnormal births. A birthing kit, which can be a box or plastic bucket, should contain:

- Soap
- Clean soft cloths
- Sharp knife or scissors to cut the umbilical cord (if necessary)
- A small jar or bottle of 7 percent iodine (Size should be such that it can be turned up against the kid's navel.)
- Tetanus/clostridium C&D vaccine and syringes
- Ear tags or tattoo kit
- Pen, pencil and paper to record birth information

PREPARATION FOR KIDDING IN INTENSIVE SYSTEMS

The best guarantee of birthing normal, healthy kid(s) is adequate nutrition and good healthcare for the doe during the 146 to 156 day gestation period.

Two weeks before parturition, prepare the kidding pen. The pen should have fresh bedding and be clean, dry and well-ventilated. Keep the birthing kit nearby and make sure that clean water can be readily accessed from the pen.

A week before the doe's expected parturition date put her in the kidding pen with plenty of bedding. If there is more than one doe ready to kid, the kidding space can be shared. Give the doe grass hay, a protein concentrate, mineral salt and water. The water bucket should be outside the pen with an opening for the doe to put her head through to drink or hang the bucket from a secure nail or hook. Check on the doe often. Change the bedding as needed to keep it dry and fresh.

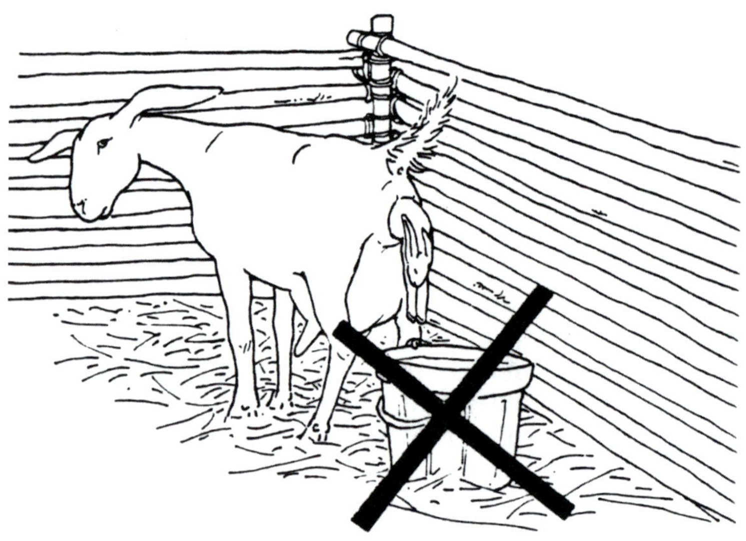

WARNING

Do not leave the water bucket in the pen. If left in the pen, the kid may drop in the water during parturition and drown.

PREPARATION FOR KIDDING IN EXTENSIVE SYSTEMS

As the expected kidding date approaches, does can be relocated from the mountains or pastures to near the main farm house. Many times, however, goats will kid in the open pasture or field, which has the risk of dogs and other predators. In either scenario, keep the birthing kit accessible. After kidding, be prepared to move from doe to doe recording dam, number of kids, birth weight and condition of kids. This is the ideal time for treating navels and for ear tagging or tattooing.

SIGNS OF PARTURITION (KIDDING)

While most goats do not need assistance in kidding, it is impossible to predict which does will need a helping hand. Therefore, as the kidding date nears, check the does regularly for any signs of parturition. Kids can be felt on the right side of the doe (the left side is the rumen). As long as the kids are moving, birth will not usually occur within 12 hours.

Watch for the following signs of parturition:

- Doe's udder begins to enlarge and fill with milk from six weeks to one week prior to kidding.
- Doe stands away from the herd.
- Doe is restless and paws the bedding.
- Doe has a clear or red discharge from the vulva.
- Tail lifts up.

- Doe appears hollow on either side of the tail.
- Doe is unusually affectionate, rubbing up against or licking the farmer's hand.
- Doe looks back at her sides and "talks."
- Some does alternate standing and lying down.

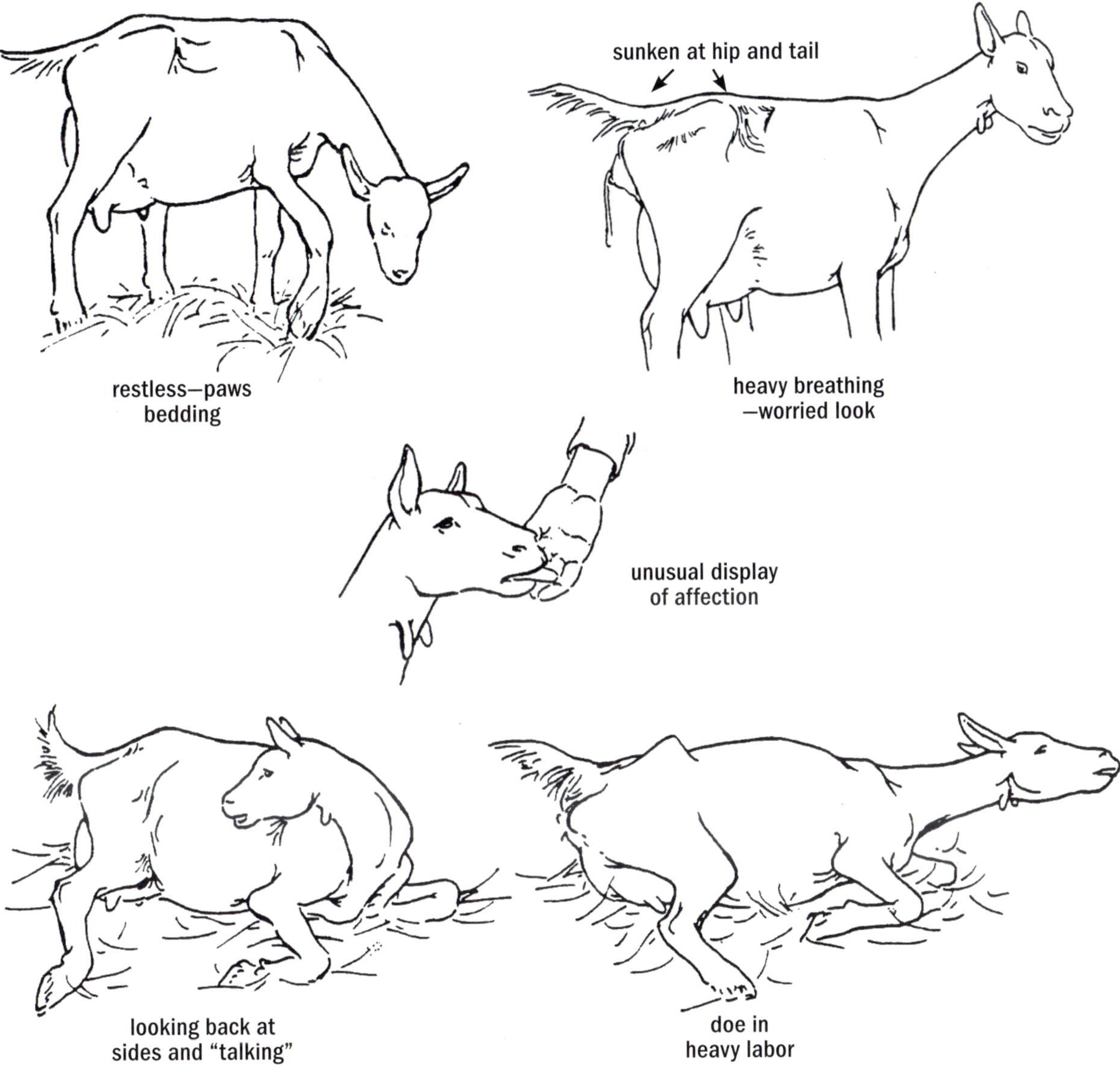

When the water sac breaks and the doe is straining hard, the last stage of labor has begun. If the water sac is not broken, you can break it for the doe. The doe may have a long discharge hanging from the vulva. The kids should be born within an hour. The doe can give birth standing up or lying down—both are normal postures. If the kids have not been born within one hour, the doe probably needs help.

NORMAL BIRTH POSITIONS

There are two normal delivery positions.

- The nose diving between the front legs. (Figure A)
- With both hind legs together and the hind dew claws up. The kid should be right side up, face down. (Figure B)

When there are twins, one will come with diving nose, the other with feet first. (Figure C) Should the two both be in the diving position, there may be difficulties as they both try to exit the birth canal at the same time.

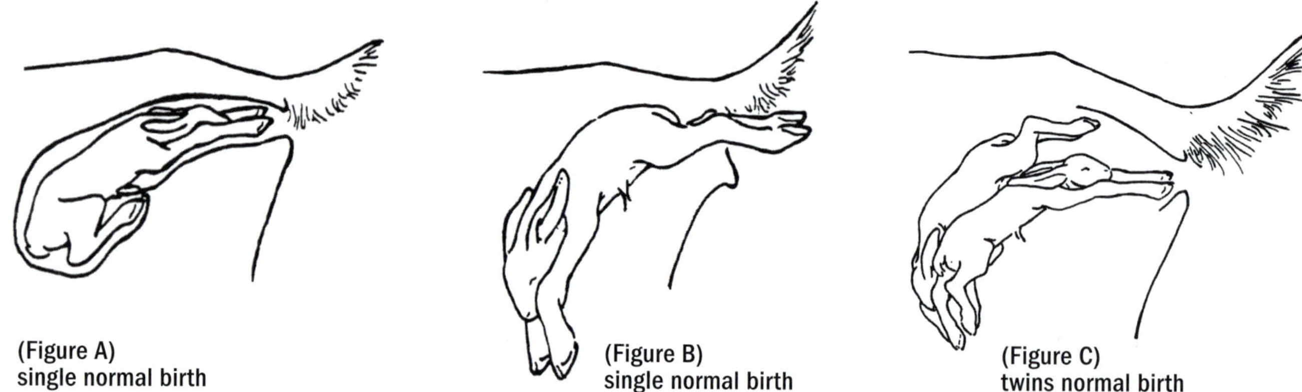

(Figure A)
single normal birth

(Figure B)
single normal birth

(Figure C)
twins normal birth

ABNORMAL BIRTH POSITIONS

- Head normal, front legs back (Figure A)
- Legs normal, but head twisted back (Figure B)
- Breech presentation (Figure C)
- Head normal, with one front leg forward, another back (Figure D)
- Kid upside down. One leg forward, one leg back (Figure E)
- Two kids, both with nose forward (Figure F)

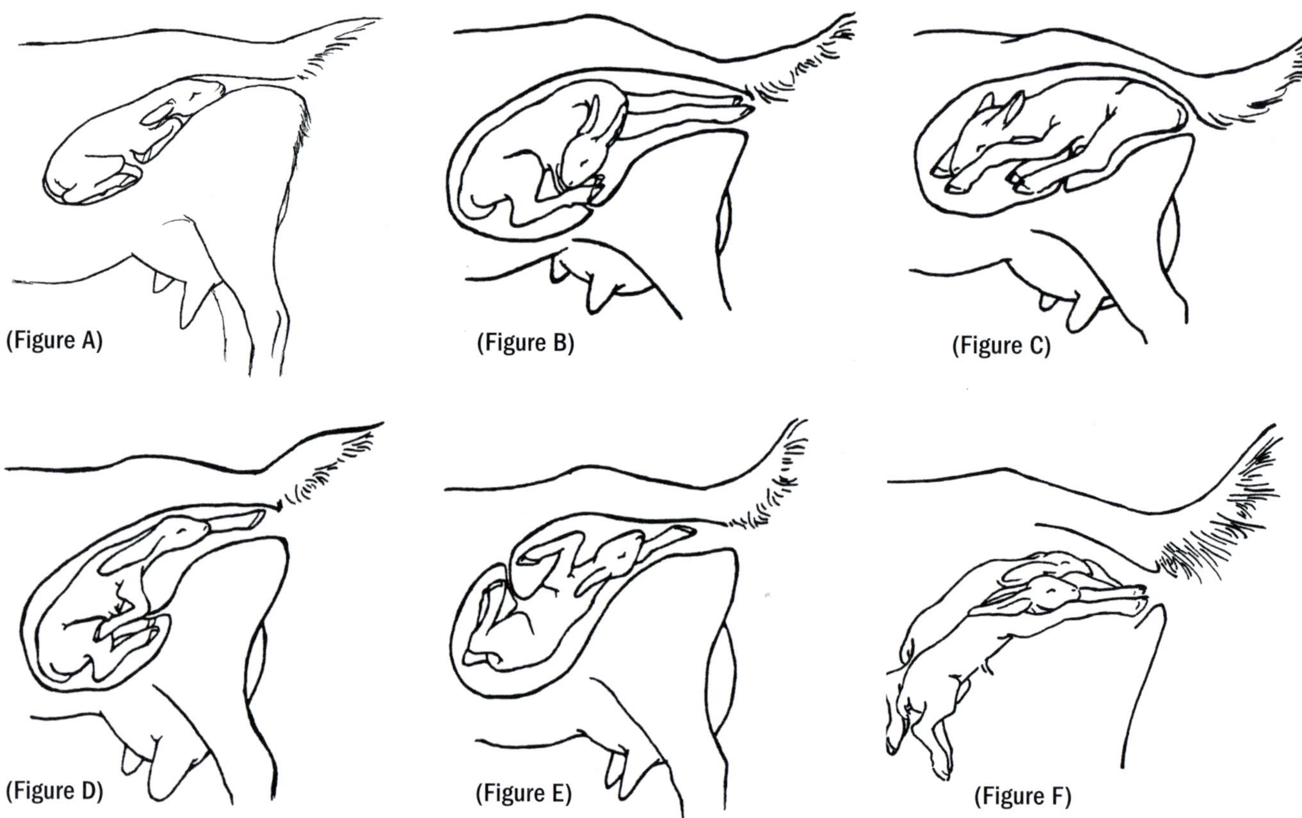

(Figure A)

(Figure B)

(Figure C)

(Figure D)

(Figure E)

(Figure F)

HELPING WITH A DIFFICULT BIRTH

If the kid is not born within one hour after hard labor begins, get experienced help to assist the doe. If the doe continues to push hard and labor seems to stop and start she may also be having trouble kidding. Parts of the kid may appear, but the doe may still be unable to deliver.

If no one is available to help, follow the instructions below to assist with the birth. People with small hands can usually reach inside the goat more easily to rearrange kids. Often women, who typically have smaller hands, are more comfortable performing this task.

Before reaching inside the doe, wash the doe's vulva with a mild solution of soap and water. If possible, have another person restrain the goat or tie her. Cut fingernails short. Remove jewelry and rings. Use warm, clean water and soap to thoroughly scrub hands and arms with soap, including the front, back, fingers, fingernails and cuticles. Remove any visible dirt. Rinse. Dry with clean cloth or air dry.

Gently insert one hand inside the doe's birth canal to determine the position of the kid(s). The doe's birth canal should feel slippery and usually no lubricant is necessary.

Move the kid until it is in a normal birth position using the following guide. (In this guide, "forward" refers to coming out of the birth canal normally, and "back" refers to moving the kid toward the uterus.)

- ***Head Normal, Front Legs Back (Figure A).*** Reach inside the doe, find the kid's neck and follow it to the kid's chest and then to the elbow of one front leg (Figure G). At times it is necessary to push the kid's body further back into the uterus to free up the legs. Hook the front leg with a finger and gently pull it forward and straight, trying to keep the hoof covered by your palm to protect the uterus (Figure H). Try the other front leg (Figure I). Then rock one shoulder and then the other gently out of the doe. Pull with the doe's contractions, not against her. Wipe off the kid's nose with a clean cloth. Clear its mouth with your finger and get it breathing. Towel the kid dry with fast strokes.

- ***Legs Normal, but Head Twisted Back (Figure B).*** This is one of the most difficult positions. Slide hand into uterus and push legs and the kid's body back into canal slowly. Cover head with palm of hand and hold head steady while bringing legs into diving position with fingers. Guide head with palm of hand and fingers until it enters the birth canal.

- ***Breech Presentation (Figure C).*** Rump first, legs deep in canal. Push the rump farther in to the uterus. Try to grab the hind legs individually, fold them up and turn them a little sideways while bringing them out towards you. Try to cover the hooves with your palm to avoid damage to the canal.

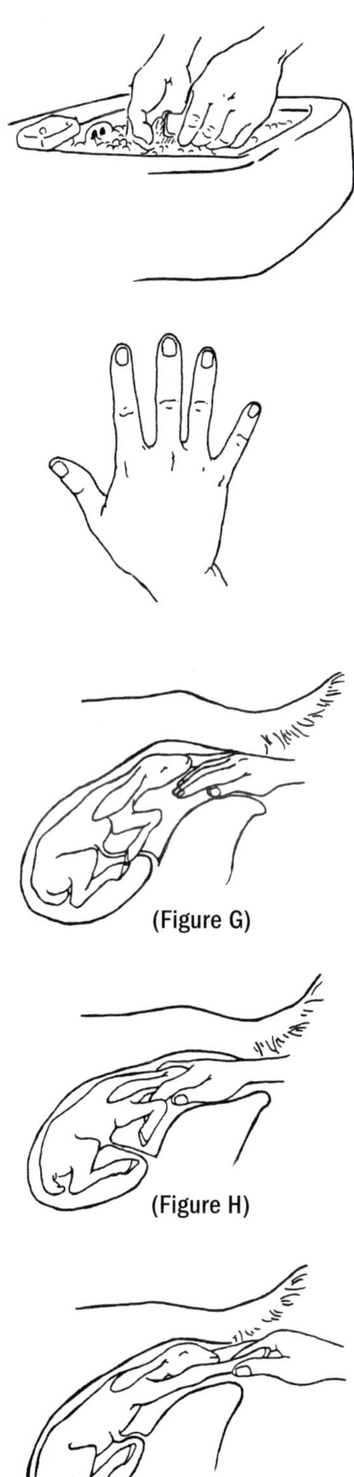
(Figure G)

(Figure H)

(Figure I)

- ***Head Normal With One Front Leg Forward and Another Back (Figure D).*** Sometimes the kid can be born in this manner. If necessary, reach in and try to grab the second front leg at the elbow, move into canal.

- ***Kid Upside Down, One Leg Forward and One Leg Back (Figure E).*** Slowly turn the kid into the normal birth position.

- ***Two Kids, Both With Nose Forward (Figure F).*** This is usually a very difficult birth. You need to carefully push one kid back into the uterus. Place the other kid in the normal birth position and pull it out. Then place second kid in birth position and gently assist.

After the doe has received assistance to safely remove the kid, enter your hand again to be sure that no other kids are left in the womb.

WARNING

Does that have had assistance during kidding are more susceptible to uterine infection. Several treatments are effective. If available, antibiotic uterine boluses can be inserted to prevent infection. An oxytetracyclin or penicillin injection can also catch an early infection. An iodine douche can be made using 2 liters of water with ¼ cup iodine. The color should resemble a weak tea.

GOOD GENERAL RULES

Within one hour of birth a kid should:

- Nurse colostrum
- Have navel dipped in 7 percent tincture of iodine
- Be identified by ear tag or tattoo
- Have date of birth, ear tag and birth weight recorded

Women of childbearing age should always wear gloves when assisting births, touching amniotic fluid or placenta. Otherwise they can get toxoplasmosis or leptospirosis which can severely damage or abort any fetus they are carrying.

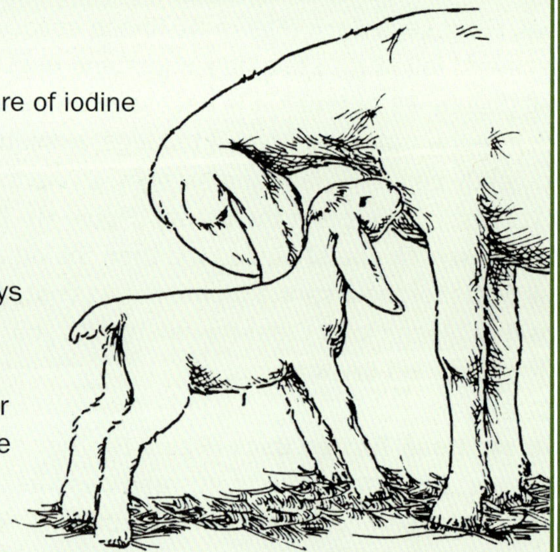

THE BIRTH PROCESS

STAGE ONE: CERVIX IS DILATING

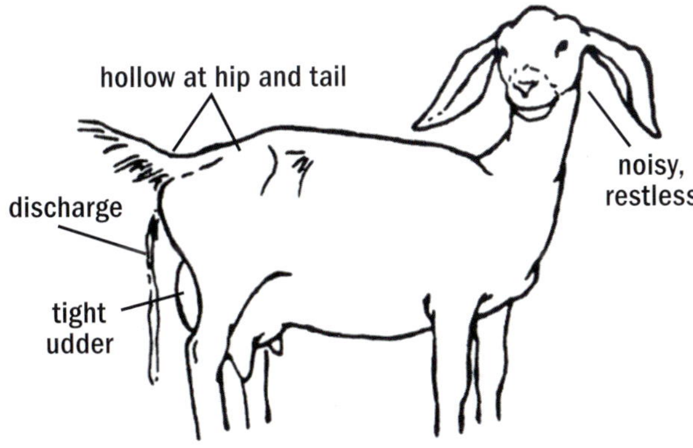

STAGE TWO: KID IS IN PASSAGE

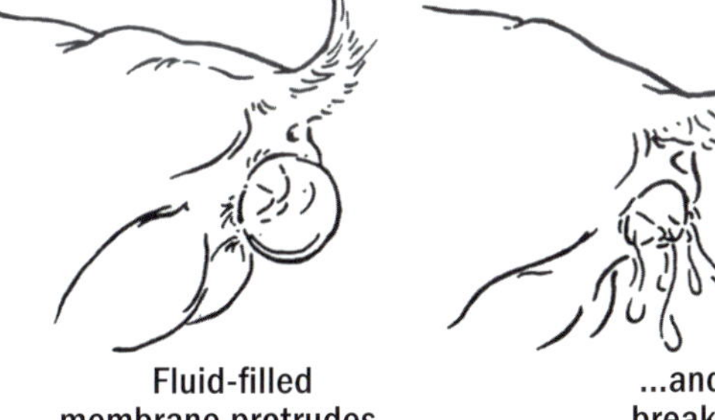
Fluid-filled membrane protrudes...

...and breaks

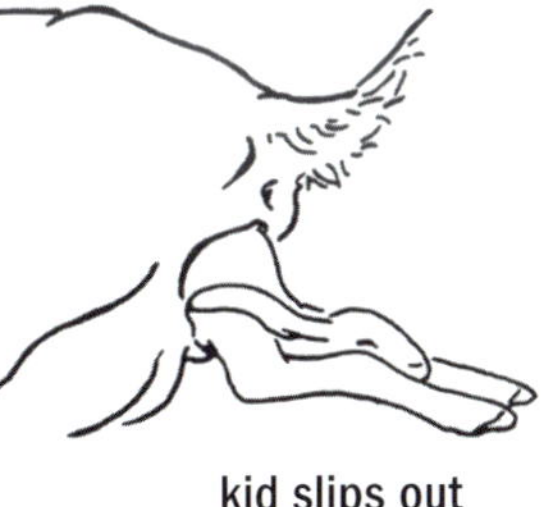
kid slips out easily now

STAGE THREE: KID IS BORN

Doe should clean kid now unless another is on the way. Finish with drying the kid with towel.

Apply iodine to the navel. Give the kid(s) colostrum as soon as possible.

AN PEISCHEL, GOATS UNLIMITED

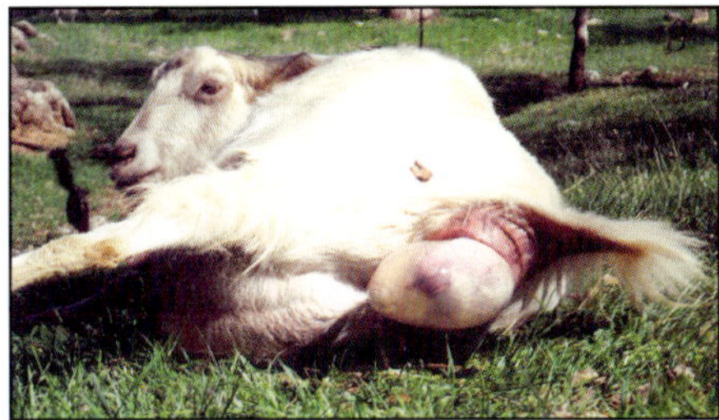
▲ *A doe strains in the beginning stages of hard labor. The nose and front feet of the kid are visible.*

▲ *The doe licks the first kid as she begins to deliver the second.*

▲ *The doe cleans the second kid. The first born starts to stand and moves toward her mother.*

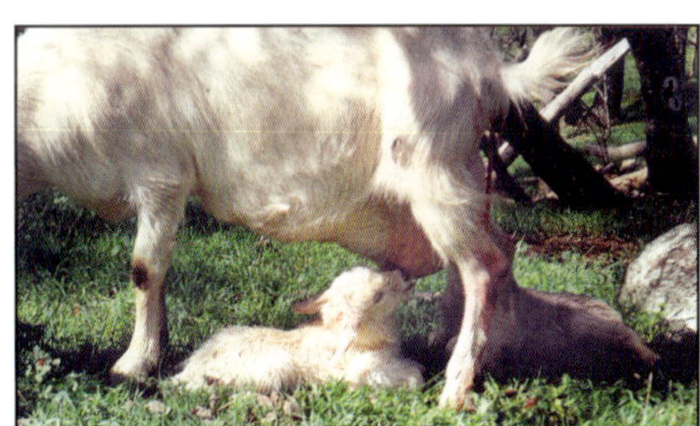
▲ *The first kid nurses, receiving colostrum which contains antibodies to protect the kid from disease.*

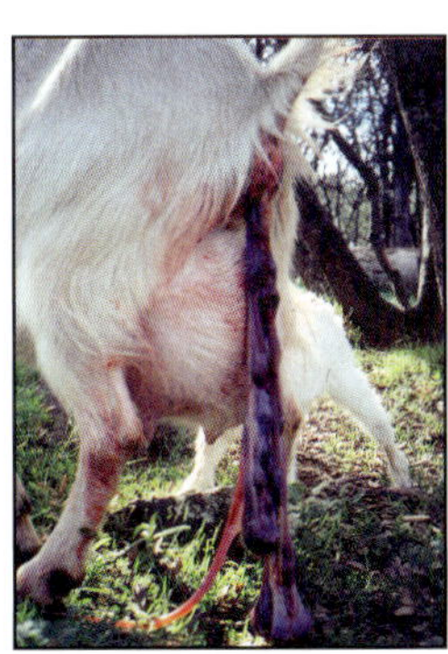
▲ *Afterbirth continues to detach and be expelled from goat's uterus.*

▲ *Some does will eat the afterbirth following delivery. Bury any remaining afterbirth so it will not attract dogs or predators.*

CARING FOR NEWBORN KIDS

If the water sac is not broken, you can break it for the doe. Clean mucous out of the kid's nose, throat and mouth. A straw up the nose provokes sneezing which helps clear the airways. If the kid appears lifeless, swing it by its hind legs—fast and hard or use fingers to give short hits the middle of the newborn's heart. Swinging the kid is normally the best solution, however, as it dislodges any mucous and increases circulation quickly.

After the kids are born, let the umbilical cord break naturally so all the blood in it flows into the kid. As soon as the kid is born, dip the umbilical cord in a 7 percent tincture of iodine solution to prevent deadly germs from moving inside the kid's body. To do so, cover the navel with a small container containing the iodine solution. Turn the opening against the kid's body and hold for one minute.

See the Feeding Newborn Kids Section in Lesson 3 for complete information on the recommended diet for kids from birth to 12 weeks and view the Resource Section on CAE Prevention.

CARING FOR GROWING KIDS

Daily observation of kids is important. Be sure they are kept warm and dry and have plenty of exercise and sunshine. Have small feeders for forage and protein concentrate available as well as mineral salt and water.

Two or three kids under one month of age that have been separated from the doe can be placed in a 1 X 1 meter box with solid sides and bedding to prevent exposure to drafts. Change the bedding every day and exercise the kids. Even when the kids are outside in the day time, keeping them in the "kid box" is a good precaution. Use this only with young kids and do not crowd.

DISBUDDING KIDS

Disbudding is not widely practiced outside of the United States. It is used to prevent goats from getting tangled in the bush or fencing and to limit aggressive behavior.

The decision to disbud or dehorn kids is a management decision. If opting to do so, disbudding should be done when goats are between three days to two weeks of age. After two weeks, the horn is too big. Dehorning older goats is not recommended as there is only a thin membrane separating the horn from the brain.

BUILDING AND USING A DISBUDDING BOX

From a Design by Gordon Mills and Jan Kelley

Materials for Building

A disbudding box makes it possible for one person to accomplish the job. Materials needed are ½" plywood, wood, nails, wood screws, a hammer, screwdriver and glue. You will also need a 2" x 8" block to cut out the hole for the head in the final part of assembling the box. Two hooks and eyes or other fasteners will be needed to secure the top of the box. Plan to put a leather handle on top of the box to make it easy to carry. The box must be sturdy enough to sit on.

- ¼" plywood/masonite for box
  2 pieces for sides (24"W x 18"H)
  2 pieces for end (6½"W x 18"H)
  1 piece for top (24"W x 7"H)
  1 piece for bottom (23½"W x 6½"H)
- 1 piece of 2" x 8" wood for head piece
  (2"W x 7"L x 6"H)
- 1"x1" wood for corner bracing
  4 pieces (16¾"L)
  2 pieces for side bottom (23½"L)
  2 pieces for end bottom (4½"L)
- 1 door hinge
- 2 hooks and fasteners
- Nails
- Wood screws
- Glue

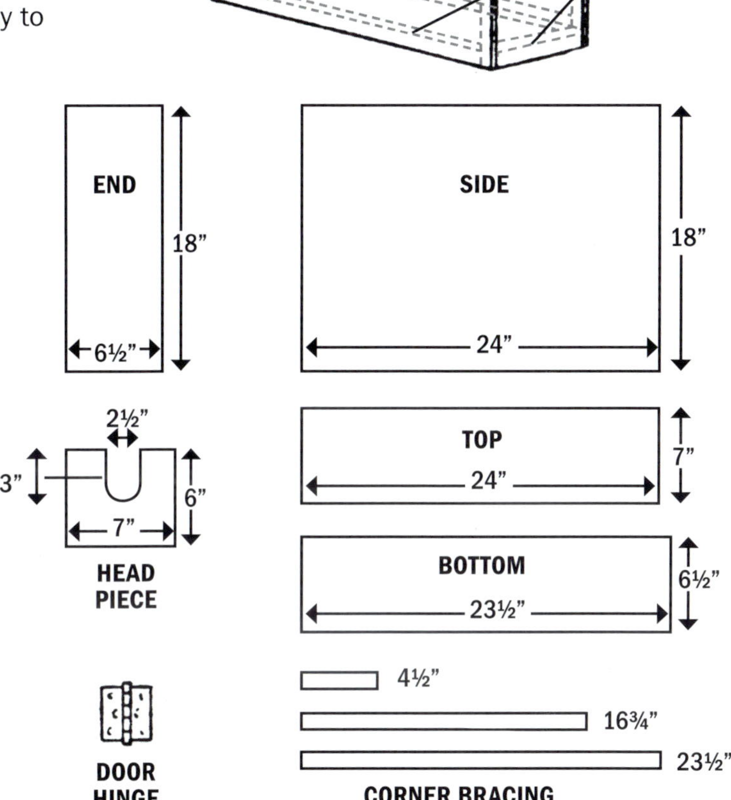

Equipment Needed for Disbudding

- A thick metal rod - about 1 cm in diameter (¾")
  (the metal rod should have a handle on it which will not conduct heat)
- A hot fire or an electric disbudding tool (see photos next page)
- A person to help or a disbudding box
- Sharp scissors for removing hair around horn buds

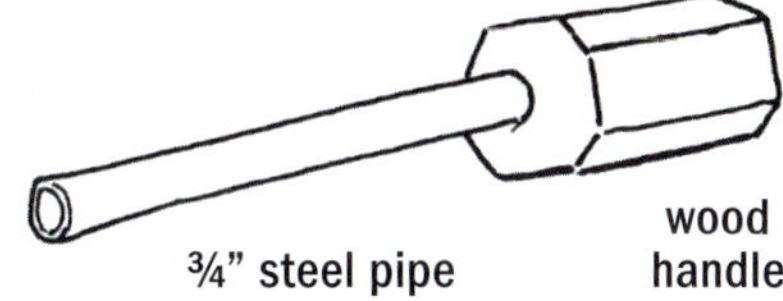

How to Disbud the Kids

Select the goat kid(s) to be disbudded and have them in a pen nearby. Be as gentle as possible with the goat during each procedure. The kids should be three days to two weeks of age.

Heat the electric disbudding iron, or a non-electric iron in a wood or charcoal fire. Place these tools on a stone within easy reach. When using a non-electric tool, it may be best to have

To disbud kids, burn area shown by dotted circle

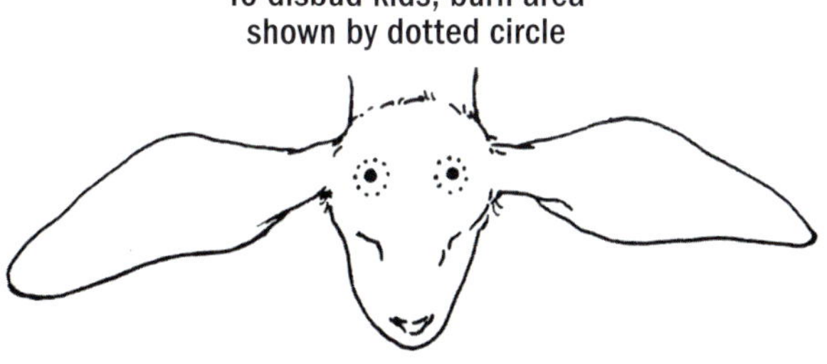

a helper who can bring the iron to you so that it stays "red hot." An electric iron should have a cord long enough to reach the box and disbud the goat. It should also be "red hot." An iron that is not hot will make it impossible to accomplish the task quickly. You will also want your disinfectant and a bottle of warm milk nearby.

(If this is your first time—practice holding the iron and turning it against a piece of wood so that you get the feel of the iron in your hand. Twist back and forth on the wood until you see a circle similar to that you will see on the goat's head.)

Place the kid in the disbudding box, straddle the box and sit on top to be directly over the goat; or have an assistant hold the kid tightly between his/her knees, exposing the head. Feel for the horn buds. Remove any excess hair with the sharp scissors. With one hand, steady the head, holding the lower jaw and muzzle. Place the hot iron firmly and directly over the horn bud. Hold firmly for 10 seconds, pushing against the head and twisting the iron. Expect the kid to cry out during this procedure.

The hotter the iron, the shorter the duration. A copper colored ring is a good sign of successful disbudding. Release the animal. Give the kid a bottle of warm milk or put the kid back with the doe.

HEIFER INTERNATIONAL

▲ *Disbudding a kid*

HEIFER INTERNATIONAL

▲ *Disbudding box and tool*

WINROCK INTERNATIONAL

▲ *Disbudded kid*

Scurs

Scurs are small portions of horn which grow back after incomplete disbudding. They are usually obvious by two to four weeks of age. Apply the disbudding iron to the edges of the scur until a copper ring is formed. The scur should fall off one week later. If scurs are cut without killing the cells underneath, it will continue to grow.

POST-NATAL CARE OF THE DOE

Keep the doe confined until the afterbirth has been completely discharged. Although some does will eat the afterbirth (placenta), be sure to dispose of any afterbirth that is not consumed. Give the doe a warm drink of water with a little molasses or sugar. Be sure the doe is eating and drinking. Observe the doe's ability to be a good mother.

Within the first hour following birth, it is important for the kids to nurse in order to receive the antibodies contained in the colostrum that the doe produces in the first 72 hours. If the doe is a heavy producer, after the kids nurse, the doe can be milked out and the colostrum can be set aside and reserved for orphan kids.

Hard, enlarged udders (udder edema) are occasionally seen when a doe kids. This is not necessarily a sign of mastitis (infection). Does with udder edema will have swollen and warm udders, but the milk will be normal. Milking the doe several times each day will relieve this condition. Udder edema does not affect the quality of the milk.

Lochia

Does may have a bloody vaginal discharge for up to two weeks after kidding. Called lochia, this discharge is normal as long as it does not smell bad and the doe is eating and walking. A uterine infection causes the discharge to look like a mixture of pus and blood, and have a foul smell. Most does with metritis (uterine infection) will not feed well and act depressed. Metritis is a serious medical condition and antibiotics (oral, injected or bolus in uterus) are indicated. Get professional assistance if the doe presents signs of Metritis.

Mastitis

Mastitis is an infection of the udder caused by bacteria and other microorganisms. Mastitis is characterized by heat, pain and swelling of the udder. Hard knots may be present in the udder along with some discoloration and abnormal qualities in the milk. Milk may be almost normal, watery and pale, dark yellow, thick, chunky, greenish or bloody. Cleanliness is the best way to prevent mastitis. A goat forced to lie in the mud and manure will probably come into contact with the organisms that cause mastitis. If the doe has mastitis, milk the doe out several times each day and discard the milk.

A veterinarian or a trained technical person can easily treat mastitis using intramammary or intravenous antibiotics. If possible have a culture of the infected milk to determine which antibiotic will be most helpful. Hot packs can also be utilized to relieve the doe's discomfort. Always consult a veterinarian or extension agent for serious cases of mastitis.

WARNING

Do not feed mastitic milk to kids or other animals. Studies have shown that female goats fed mastitic milk are more likely to have mastitis themselves. After treatment for mastitis, wait three to four days before milking for human consumption.

CYCLE OF LACTATION AND DRYING OFF THE DOE

Good nutrition is the single most important thing for the milking doe. If she has adequate energy and protein in her feed, plenty of clean, fresh water, proper health care and exercise she can provide milk for up to 300 days a year. The length of lactation will vary according to environment and feed resources. If given rice bran, wheat bran, cottonseed cake or other protein supplements, a dairy goat will give more milk.

Stop milking the doe when she is three months pregnant. Her udder will become hard, but gradually the milk will stop and she will not produce any more. If she is too uncomfortable, she can be milked out again. Be sure to milk her out completely each time. Skip a few days between each milk out.

It is important to dry off the doe (meaning to stop milking) during the last two months of pregnancy because the kids are growing fast inside the doe and need the sustenance her body provides. Milk production competes for that sustenance.

Gaining More Than Money

▲ Wu Yuantian feeding his goats.

Wu Yuantian is a farmer in Weijiaba Village of China's Lezhi County who raises crops and animals. His family lived with an annual income in 2005 of US$212. The family wanted to improve their livelihood, but without the necessary skills and startup capital, this was difficult.

In January 2006, members of the Lezhi Promotion Association for Rural Development came to Wu Yuantian's village to explain Heifer's program. Wu was the first villager to sign up. He was very excited hearing how Heifer China helped other needy communities move out of poverty and closer to self-reliance. After discussion with his family, he applied for the project the following day. The project staff approved his application and after training, in June he received 10 Lezhi Black goats.

The Lezhi Black Goat is a domestic breed that was improved by selective breeding of local black goats in the 1990's. Farmers favor the Lezhi Black Goat because it is an improved meat breed and has high productivity.

Wu's family treated the goats like members of the family. Wu and his family actively participated in all project activities and strictly followed the goat management requirements introduced by the Lezhi Promotion Association for Rural Development.

Their hard work paid off. Now they have 23 goats, including 15 kids. Eight does have gained a total weight of 354 lb. Wu and his family sold two goats for US$76 each. At that rate, they estimate they can make US$622 from goat raising in the first half of 2007.

The goat manure from their small herd has provided more fertilizer for cropland, saving them the US$50 they might otherwise need to pay to buy manure. The goat operation has also generated bio-gas for cooking and light, saving firewood and

Continue on page 48

Continued from page 97

▲ *Wu Yuantian and his wife chopping forages to feed their goats.*

contributing to local environmental conservation. And now the family has more than tripled its meat consumption. Previously, Wu's family ate meat once every 10 days. Now they have meat once every three days.

Before they participated in the project, during the slow season, Wu went fishing or simply hung around in town, leaving his wife, Wen Yongqing, home to do the housework. Now, because the whole family is dedicated to raising goats, Wu and his wife share responsibilities. When one is busy, the other takes over.

They have agreed that Wen keeps the income gained from selling goats, and whenever a family member needs money, they discuss it together. Relationships among family members are continually improving.

Relations with neighbors have also improved. Since Wu was the project's earliest participant, neighbors often come to learn from him. Yu Yingsi is one of them. "Wu Yuantian is hard working, experienced and warmhearted," Yu said. "I learn a lot from him about goat management and caring for kids."

Asked about his future in the project, Wu said, "I'm happy to participate in this project. It gives me more than money. If I keep working hard, I will complete passing on the gift later this year and renovate my house in three years." ◆

THE LEARNING GUIDE

MILKING THE DOE

LEARNING OBJECTIVES

By the end of the session, participants will be able to:

- Identify milking equipment and supplies needed for a successful goat dairy
- Milk a doe
- Pasteurize milk
- Complete a daily milk record chart
- Make fresh cheese

TERMS TO KNOW

- Lactation
- Pasteurize
- Udder
- Teats
- Teat dip
- Mastitis
- Udder edema

MATERIALS

- At least four lactating does
- Milking stand with place for feed
- Feed for doe
- Teat dip
- Teat dip cups
- Bucket of warm soapy water for washing hands
- Bucket of water for rinsing
- Bucket of drinking water for doe
- Clean cloths
- Milk bucket
- Strip cup for examining milk
- Milk filter or straining cloth
- Daily milk record sheet
- Pan for pasteurizing
- Pan for making cheese
- Fire or stove for pasteurizing
- Lemon juice or vinegar
- Salt and herbs for cheese
- Milk thermometer (optional)
- Wooden spoon
- Straining cloth for cheese
- String for tying cloth holding cheese and hanging

ADVANCE ASSIGNMENT

Locate an active goat dairy where the session can be held. Check with farmer about what equipment and supplies will be needed as per the above list. Assign two class members to gather herbs for cheese making.

TIME (May vary according to group)	ACTIVITIES
15 minutes	**Group Sharing** Ask the group to share a fun thing they did since the last session. (This is for a change of pace.)
30 minutes	**Get Everyone Thinking and Talking** ■ Divide participants into two groups. ■ Ask each group to create a five minute skit about "Why Milk is Good for You." ■ Groups will present their skits. ■ Other participants are invited to make comments or ask questions.
60 minutes	**Practice Proper Milking Procedure** ■ Demonstrate proper milking procedure including checking milk in strip cup for abnormal signs, using teat dip and recording production on a daily milk record.
15 minutes	**Prepare Milk for Pasteurizing** ■ Demonstrate methods of filtering milk. ■ Pasteurize Milk and determine cooling procedures.
20 minutes	**Make Fresh Cheese** (See appendix for recipe and directions.)

REVIEW-20 MINUTES

- What was useful?
- What was a surprise?
- What did not work out well?
- What do you know now that you did not know before?
- What can you do as soon as you get home?
- What practices will be difficult to do at home?

THE LESSON

MILKING THE DOE

Healthy dairy goats provide a continuous source of nutritious milk. With the proper milking equipment and procedures, milk, cheese and other dairy products can be produced not only for personal consumption, but also for sale, generating an additional source of income.

MILKING EQUIPMENT AND SUPPLIES

The following equipment is recommended for beginning small-scale milk producers:

- A milking stand (optional, but recommended)
- A four to eight liter (1-2 gallons) stainless steel or enamel milk bucket*
- Two cups (one for examining the milk prior to milking and one for teat dip)
- A filter, strainer or a clean cloth for straining milk
- A pan for pasteurizing milk
- A container for storing the milk. Avoid plastic as it is difficult to sanitize
- A bucket of cold water to cool milk
- Teat dip**
- Soap
- Water
- Clean cloths
- A soft brush to loosen dirt and hair from udder (if needed)

Be sure all of the milking equipment is kept clean. To do so, wash equipment with soap and warm water. Rinse well. Turn upside on a clean cloth and air or sun dry. You can dry on a rack outside in the sun as well as in the house. Cover the equipment with a clean cloth when not in use.

* *Plastic is porous and holds bacteria and micro organisms, therefore stainless steel or enamel buckets are recommended. If only plastic is available, be sure to wash and rinse the pail well and sterilize with boiling water.*

** *Teat dip can be commercially procured or homemade. For the latter use a solution comprised of 0.5 percent iodine and/ or make a solution that is 5 percent bleach and 95 percent clean water. (Note: The USDA does not approve bleach solution as a teat dip, but in developing countries it may be the only option.)*

PROPER MILKING PROCEDURES

Does should be milked twice daily (morning and evening) on a regular schedule. Setting milking times twelve hours apart is ideal. Milk at approximately the same time each morning and evening (or late afternoon). Although some people milk only once a day, the highest milk production can be obtained through two daily milkings. Record time of milking and milk production for each doe, using the chart in the record section of this manual. To begin a milking routine follow the steps below:

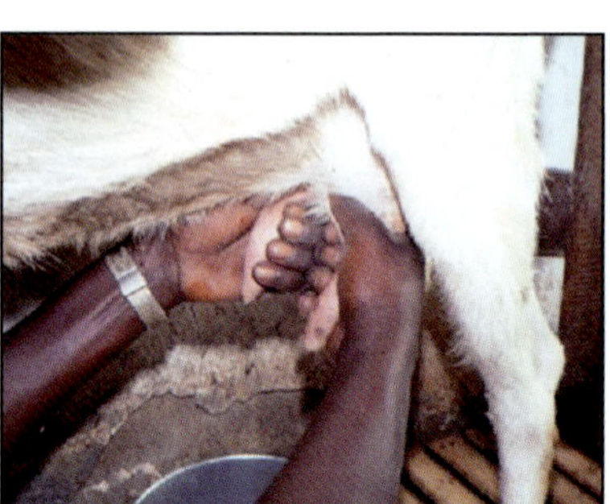
ROSALEE SINN

▲ *Milking a goat in Cameroon*

- Place feed in a bucket or in the feed container attached to the milking stand. Bring the goat to the milking location.
- Put her on the milking stand (if available) or tie her so she can eat from the bucket.
- Wash hands with soap, rinse and dry.
- Begin the milking process by visually examining the goat's udder. If the udder is dirty, brush away loose dirt and if necessary wash the udder with soap and water. If the udder requires washing dry with a clean towel. (Thorough drying of the udder is essential as contaminated water can collect at the end of the teats.)

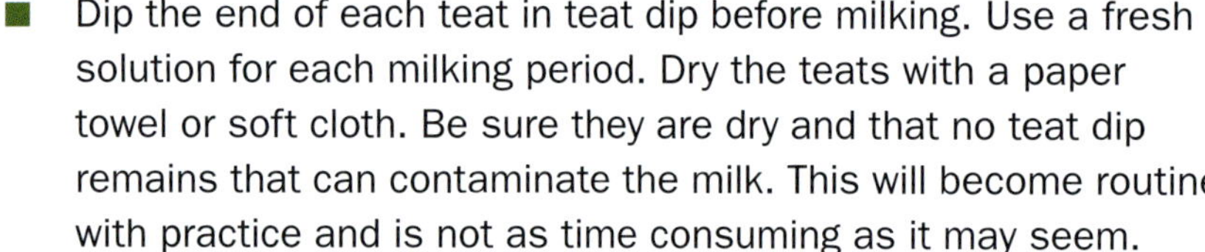

- Dip the end of each teat in teat dip before milking. Use a fresh solution for each milking period. Dry the teats with a paper towel or soft cloth. Be sure they are dry and that no teat dip remains that can contaminate the milk. This will become routine with practice and is not as time consuming as it may seem.
- Draw a small portion of milk from each teat and direct into a cup. This first squirt clears bacteria in the teat canal and allows for careful examination of the milk. If color and consistency are normal, discard milk in the cup, place the pail or other container under the goat and continue milking.
- Sit on a stool or on the milking stand next to the goat. Place both hands on the teats of the goat. Grasp the teat at the top with thumb and index finger together to trap the milk in the teat. Gently, but firmly, bring pressure on the teat with the index finger forcing the milk down to the sphincter. The middle finger does the same and then the little finger. Do not drag (pull) or jerk down on the udder. Use steady pressure and loosen grip before squeezing again. It should take approximately five minutes to milk the doe. When the flow seems to stop, massage the udder and express the final milk.

> **WARNING**
> **If milk is stringy, lumpy or contains blood, this is an indication of mastitis or another illness. If this occurs, do not use the milk. Milk out the udder into a separate container and discard. (See Lesson 6 for information on Mastitis and Lesson 8 for proper treatment.)**

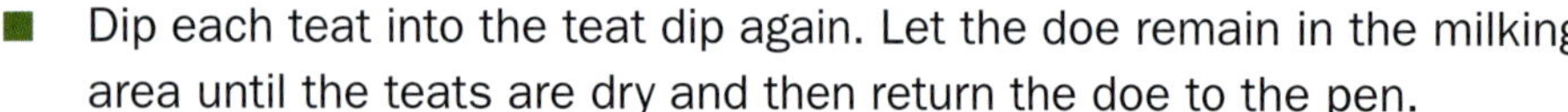

- Dip each teat into the teat dip again. Let the doe remain in the milking area until the teats are dry and then return the doe to the pen.

PROPER MILKING PROCEDURES

Washing and cleanliness prevent disease.

a) Wash hands with soap

b) Washing the doe's udder before milking

c) Massaging udder for more milk

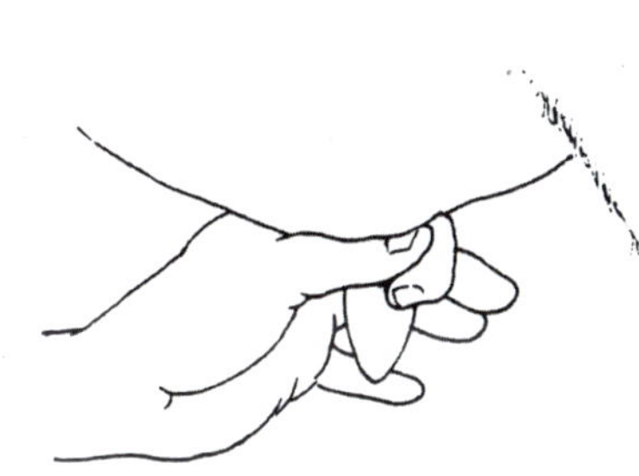

d) Trapping milk in teat by clamping at top with thumb and index finger

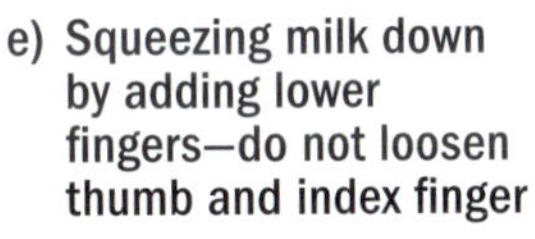

e) Squeezing milk down by adding lower fingers—do not loosen thumb and index finger

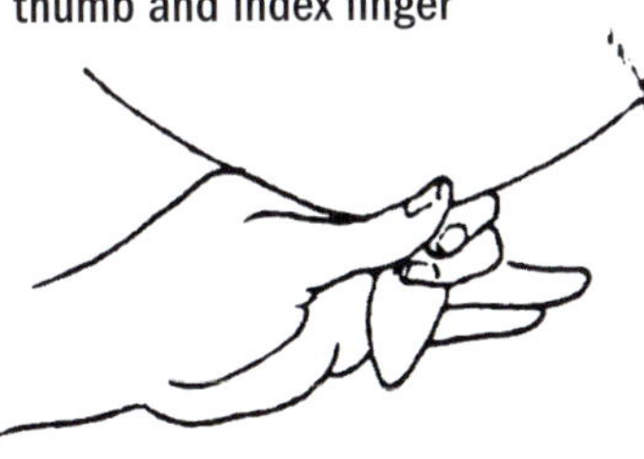

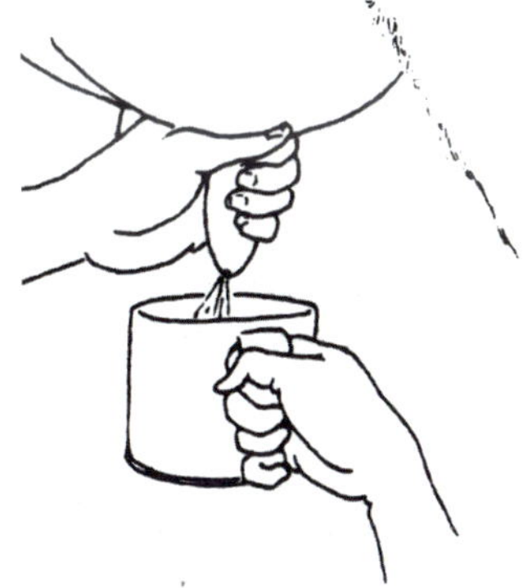

f) All fingers squeezing first squirt into the test cup

g) Dipping teats after milking

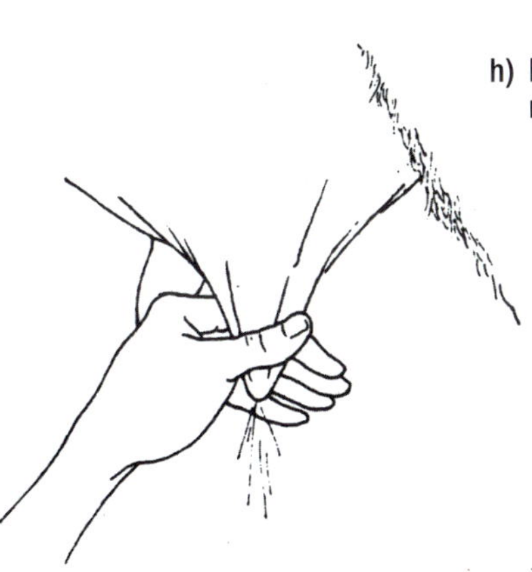

h) Do **NOT** milk like this

HOW TO CARE FOR THE MILK

Strain the milk through a filter or cloth into a clean container to remove any hair or other stray dirt that may appear during the milking process, Use commercial filters only once. Cloths can be used again if washed with soap and water, rinsed once with warm water and then with boiling water and dried in the sun.

Pasteurize the milk to kill germs and prevent transmission of zoonotic disease to kids or humans. Another advantage is that pasteurized milk normally remains sweet for several days even without refrigeration.

To pasteurize place the milk on the stove or cooking fire and bring it almost to the boiling point. Hold for one minute. If a thermometer is available, more precise treatment can be achieved by using the guidelines below:

CELSIUS	FAHRENHEIT	DURATION
63°C	145°F	30 minutes
72°C	161°F	15 seconds
100°C	212°F	0.01 second (boiling)

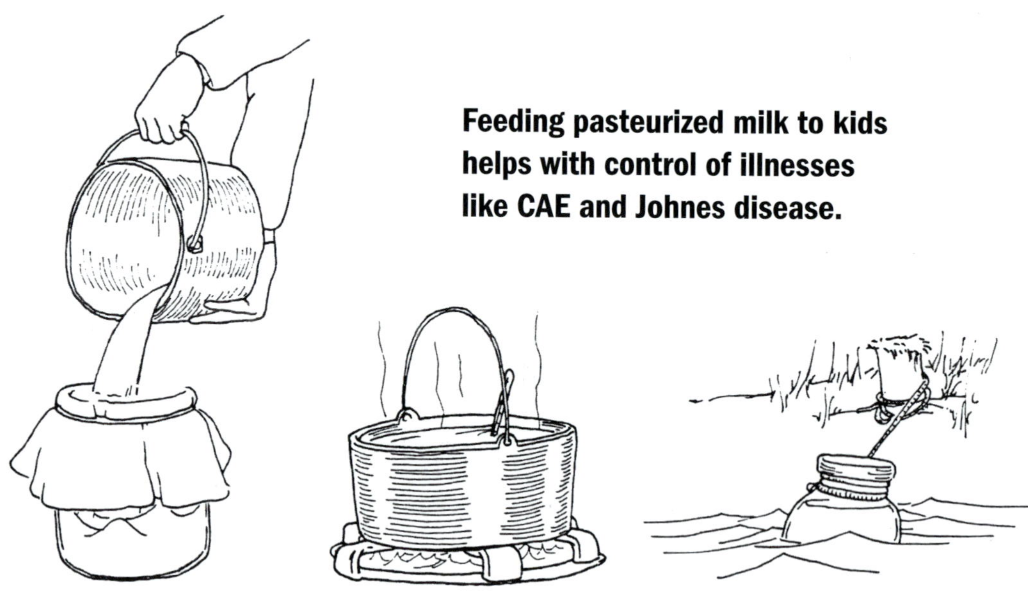

Feeding pasteurized milk to kids helps with control of illnesses like CAE and Johnes disease.

Immediately after straining, pour the milk into a clean container that has been rinsed with boiling water. A jar with a tight cover is ideal. Place the container in a bucket of cold water or in a stream to cool. If a refrigerator is available, the milk will be usable for at least a week. Sour milk can be used for cooking.

If possible have a veterinarian test does for Brucellosis and Tuberculosis (TB) each year. These are zoonotic diseases that can be passed on to humans through consumption of dairy products.

INDIVIDUAL DAILY MILK RECORD SHEET

Name of Goat:____________________ No:__________ Year:__________

Weigh milk and record in kilos or lbs. If no scale, measure after straining and record in liters or quarts.

DAY	JAN		FEB		MAR		APR		MAY		JUN		JUL		AUG		SEP		OCT		NOV		DEC		TOTAL
	am	pm	am	pm	am	pm	am	pm	am	pm	am	pm	am	pm	am	pm	am	pm	am	pm	am	pm	am	pm	
1																									
2																									
3																									
4																									
5																									
6																									
7																									
8																									
9																									
10																									
11																									
12																									
13																									
14																									
15																									
16																									
17																									
18																									
19																									
20																									
21																									
22																									
23																									
24																									
25																									
26																									
27																									
28																									
29																									
30																									
31																									
TOTAL																									

Building Opportunity

The Comiza family lives in the remote and predominantly Roma village of Nemsa in central Romania, far from the opportunities that larger cities offer and without public transportation.

Iulius Comiza, the father, suffers from tuberculosis and cannot work. His wife, Dorina, is a housekeeper. The couple has two daughters, Madalina, 14, and Diana, 21, and two sons, Darius, 10 and Flavius, 17. Diana and Flavius have left the village to find work in larger cities.

The family lives under very difficult conditions because jobs in the village are almost non-existent. About 80 percent of the villagers are unemployed. The Comiza family's income sources are Iulius' disability pension and the underage children's state allocation. All these add up to a meager income, barely enough for basic needs such as food, gas, electricity, clothing, school supplies, medicines and transportation.

In 2004, Heifer Romania started the Roma People's Goat Project in Nemsa. Since then, Iulius has been one of the most enthusiastic and dedicated among the Roma. He was the first to gather people together for training sessions and the first to greet visiting Heifer representatives in Nemsa. Despite his illness, he has joyfully participated in Heifer meetings all over the country.

Iulius decided to receive goats only after the poorest Roma families have received their share of animals. In 2005, the Comiza family received their first two passed-on kids. Another project beneficiary, an old woman in the village, fell ill and couldn't care for her three Heifer goats anymore. Iulius offered to adopt them, and in addition the family acquired eight other goats. Today the Comiza family is the proud owner of 13 goats. Each family member helps take care of the goats, feeding them hay, corn, barley and sometimes beetroots or potatoes. Darius is particularly fond of the animals.

▲ *Darius with his goat.*

The goats have improved the economic and nutritional stability for this Roma family. Milk and goat cheese have improved the family's nutrition. In addition the Comizas sell some of the goat milk for income. Currently, each of the Nemsa goats produces an average of about 2 liters of milk a day. With milk selling at about 33 cents a liter, the Comiza family can earn as much as US$227 a month, the equivalent of a low-to-average income in Romania.

In light of the promising developments in Nemsa, Madalina has decided to stay with her family and to acquire goat-breeding skills, instead of following her siblings to the city to look for work.

The Comiza family's success is a reflection of the success enjoyed by the entire village. In fall 2006, Heifer Romania field assistants successfully crossbred 105 Carpathian goats in Nemsa with Saanen goats through artificial insemination. Some 75 crossbred kid goats have been born since then. The crossbred kids are larger and stronger than the Carpathian goats and will surely give more milk as they get older.

Heifer has helped the Roma in Nemsa build a milk collecting center where goat milk will soon be cooled and stored under hygienic conditions before being sold to larger milk processing plants. In addition to milk sales, the villagers hope to earn money by selling kid goats. Using goat manure to fertilize gardens will yield even more benefits. ◆

THE LEARNING GUIDE

HEALTH CARE FOR YOUR GOATS

Health Care and Disease Prevention with a Guide to Disease Diagnosis and Treatment

LEARNING OBJECTIVES

By the end of the session, participants will be able to:

- Identify a healthy goat by observation
- Identify signs of a sick goat
- Describe parasite prevention and treatment
- Take a goat's temperature and determine if it is abnormal
- Safely give a goat a drench, a bolus and a vaccination
- Use the health charts and know when to call a veterinarian

TERMS TO KNOW

- Drug resistance
- Liver fluke
- Life cycle
- FAMACHA chart
- Withdrawal time
- Insecticide
- Internal parasites
- External parasites
- Intramuscular
- Subcutaneous
- Infection
- Vaccination
- Trauma
- Dewormer
- Sanitation
- Supportive care
- Zoonotic

MATERIALS

- Goats of varying ages and conditions*
- A FAMACHA Chart (if available)**
- Observation Sheets from this text
- Health Charts from this text
- Thermometer (digital is preferred)
- Tube for administering bolus, and appropriate bolus medication
- Bottle or appropriate size syringe for drenching
- Syringes and needles to practice SubQ and IM injections
- Clostridium perfringens type C and D vaccine

ADVANCE ASSIGNMENT

This lesson is best taught on a farm. This activity needs to be planned in advance in order to have suitable goats for the activities. Print Observation Sheets and Health Charts for each class member to use.

* *If the group is working with goats that are ill, be sure to disinfect hands and shoes upon leaving farm.*

***To receive information about the FAMACHA Chart, e-mail famacha@vet.uga.edu*

TIME (May vary according to group)	ACTIVITIES
15 minutes	**Group Sharing** ■ Any questions from last session? ■ What are some of the common illnesses that we face as human beings? ■ Are there any diseases that are similar between humans and goats?
30 minutes	**Get Everyone Thinking and Talking** ■ Ask participants: • How can you tell when your animals are sick? • How does a sick animal impact the other parts of your life? ■ Hand out Observation Sheets and have each member of the group fill out an Observation Sheet on one of the goats that are present. ■ Discuss the results: What signs were found? What are the common causes?
20 minutes	**Observe the Goats for Internal and External Parasites** ■ Observe the mucosa of several goats to check for anemia, using the FAMACHA Chart (if available). Otherwise discuss whether the mucosa around the eyes or the goat's gums are pale or pink. ■ Check fecal samples if possible. ■ Discuss parasite problems in goats and list four points on how to prevent them.
20 minutes	**Taking a Goat's Temperature** ■ Take a goat's temperature. ■ Determine if it is in the normal range.
20 minutes	**Practice Drenching and Giving a Bolus**
20 minutes	**Practice Vaccinating Goats** ■ Vaccinate one or more goats. ■ Make notation in the goat's record.

REVIEW-20 MINUTES

- What was useful?
- What was a surprise?
- What did not work out well?
- What do you know now that you did not know before?
- What can you do as soon as you get home?
- What practices will be difficult to do at home?

THE LESSON

HEALTH CARE FOR YOUR GOATS

Health Care and Disease Prevention with a Guide to Disease Diagnosis and Treatment

Good general care and preventive health practices are essential to a successful goat enterprise. Unwanted disease can curb growth, lower productivity and ultimately result in the death of a valuable asset. Moreover, it is more economical to prevent disease than to treat illnesses. Approach goats as individuals. Be aware of any changes in the environment. Always be gentle. Animals need sunlight, fresh air and exercise, but need protection from too much sun, rain and wind. Give adequate nutrition: energy, protein, vitamins, minerals and water. Pay attention to sanitation: clean pens, feeders and water buckets will go a long way to keeping your herd healthy. Provide supportive care when an animal seems unhealthy and separate sick animals from the herd immediately.

PRACTICES FOR GOOD HEALTH CARE

- Daily observation
- Good daily management practices
- Balanced nutrition, which includes forage, protein and energy feeds, salt, minerals and plenty of clean water daily
- Measures to prevent internal and external parasites
- Good sanitation
- Clean pens, feeders and water buckets
- Appropriate shelters
- Exercise, sunlight and fresh air
- Hoof trimming as needed
- A vaccination program for enterotoxaemia, tetanus and other diseases which may be a problem in your area

SIGNS OF A HEALTHY GOAT

- Eats well and chews cud
- Has a shiny coat
- Is free of disease
- Has strong legs and feet
- Is sociable and alert
- Has eyes that are bright and clear
- Has no discharge from eyes, nose, mouth
- Has firm fecal pellets (droppings)

NORMAL PHYSIOLOGICAL MEASURES OF A HEALTHY GOAT

Temperature	38.6°C to 40.0°C or 101.5°F to 104°F
Heart (pulse) rate	70 to 80 per minute, faster for kids
Respiration rate	12 to 15 per minute, faster for kids
Rumen movements	1 to 1.5 per minute
Onset of heat (estrus)	6 to 12 months of age
Length of heat	12 to 72 hours
Heat cycle	18 to 22 days – average 21 days
Length of pregnancy	145 to 156 days – average 150 days

WARNING SIGNS OF DISEASE IN MATURE GOATS

- Not eating/no sign of cud chewing
- Grinding of teeth
- Standing off from other goats
- Weight loss
- Hair loss or rough hair coat
- Cloudy eyes (corneal opacity) or blindness
- Limping or unwillingness to stand
- A change of behavior such as moving in circles
- Dehydration*
- Fever (temperature above 40°C or 104°F)
- Anemia: pale mucosa around eyes and in mouth
- Diarrhea
- Discharge from nose or mouth
- Frequent coughing
- Abnormal swelling at any point on body
- Clots or blood in milk

* *To test for dehydration, pull skin from the goat's neck near the front shoulder area. If skin sticks to itself and does not slip back easily, the animal is dehydrated.*

COMMON SIGNS OF ILLNESS IN GOAT KIDS

Illness in kids can be more dangerous because without early treatment diseases tend to progress more quickly and result in mortality. It is therefore essential that special attention be given to the daily observation of kids, and supportive care should begin as soon as any abnormality is noted.

- Fever
- Kid does not eat
- Weakness or unable to rise or walk
- Diarrhea
- Mucous discharge from nose or eyes
- Heavy breathing
- Dehydration
- Teeth grinding
- Swollen joints or navel

PREVENTIVE HEALTH PROGRAM

Good health care always begins with daily observation. Preventive health care programs encompass good husbandry practices, routine vaccination and deworming plans, and regularly scheduled grooming to keep goats in optimal health. Goats have specific health needs and disease susceptibilities at each age and stage of development. Therefore, when implementing a preventive health program, each age group should treated differently.

Two general rules for preventive health programs should be applied regardless of the age or stage of development. First, new animals should always be vaccinated and kept in quarantine for at least one week before being placed with other goats in the herd. Second, separate any sick animals from the herd immediately.

The following sections present preventive healthcare checklists for kids, does and bucks.

Kids

- **Newborn Kids**

  After the kids are born, let the umbilical cord break naturally so all the blood in it flows into the kid. Dip umbilical cord in a 7 percent iodine solution to prevent deadly germs from moving inside the kid's body. To do this, cover the navel with a small container of the iodine solution, turning the opening against the kid's body and hold for one minute.

  Clean mucous out of kid's nose, throat and mouth. Get the kid up and moving. Be sure the kid nurses and gets colostrum within one hour of birth. It is the only food a newborn kid should have for the first two to three days. The colostrum contains antibodies produced by the doe's immune system that protect the kid from infectious disease.

  Kids – Four Weeks to Three Months of Age

  Young kids are very susceptible to internal and external parasites, which deprive them of vital nutrition for their development. If internal parasites are a problem, a deworming program should begin for kids as young as one month of age. Prepare and read fecal samples from goats to determine if deworming is needed. Avoid excessive deworming, which may cause drug resistance.

If diarrhea is a problem, the cause may be a parasite called coccidia which mainly affects young goats. Coccidiosis is a manageable disease. It can be controlled by reducing crowding and improving sanitation through frequent removal of manure and soiled bedding. If necessary, use coccidiostats, such as decoquinate, monensin, lasalocid, amprolium, sulfamethazine or sulfaquinoxaline to treat animals. Follow the directions listed on these medications.

Treat for external parasites as necessary in areas where ticks or lice are a problem. Herbal sprays can be made from neem leaves or seeds, as well as from other plants. In severe problem areas, it may be necessary to spray with insecticide as often as every two to three weeks. Ivermectin injections treat most internal and external parasites.

- **Vaccination Program for Young Kids**

  Tetanus Toxoid

  Vaccinate twice at four week intervals. After initial dose, give mature goats an annual booster.

  Clostridium Perfringens Types C & D Bacterin

  Offers protection against enterotoxaemia (overeating disease). This is often combined with Tetanus Toxoid. Vaccinate kids at age 6 to 8 weeks and again at 12 weeks if the dam has been vaccinated. Begin at age 4 weeks if the dam was not vaccinated prior to kidding. Repeat enterotoxaemia vaccination every six months in normal conditions and every four months in problem areas.

  Contagious Ecthyma (Soremouth)

  A veterinarian or technician should administer this vaccine because the disease can be transmitted to humans. This vaccine should only be administered if this disease is a problem in the herd. The vaccination should be administered under the tail or on bare skin under front leg at point of elbow. Check in three days for a scab where the vaccine was administered. A scab is a sign that the vaccination has taken. In problem herds, each year vaccinate all new additions and kids. Re-vaccinate every two years in heavily infected areas.

  Selenium/Vitamin E

  Injections can be given at one week of age, one month and three months in areas where selenium deficiency (which causes white-muscle disease) is a problem. However, if selenium is supplemented in mineral or feed, only vitamin E injections should be given.

Pregnant Does and Dry Does

- Monitor mature goats by body condition scoring
- Perform fecal exams to determine parasites loads
- Observe the mucosa for anemia regularly
- Vaccinate for Clostridium Perfringens C & D and Tetanus four to six weeks before kidding (repeat in 6 months)
- Deworm does two to three weeks before kidding if internal parasites are a problem

- Trim hooves as needed
- If selenium deficiency is a problem in the area, give vitamin E and selenium 60 days before kidding. Be careful not to overdose.
- If a problem in your area, vaccinate for rabies, leptospirosis and other abortion-causing diseases

Bucks

- Monitor mature bucks by body condition scoring
- Perform fecal exams to determine parasite loads
- Vaccinate for Clostridium Perfringens Type C and D, and for tetanus, (plus rabies and leptospirosis if a problem in the area) on the same schedule as pregnant does
- Treat urine scald as needed with petrolatum (Vaseline, petroleum jelly)
- To prevent urolithiasis (blockage of the urinary canal) provide clean, fresh drinking water at all times. Provide salt at all times and keep a calcium: phosphorus ratio of 2:1 in the buck's diet.
- Exercise

DIAGNOSIS AND TREATMENT

Caring for a sick animal involves both supportive care and specific treatment. Supportive care is directed towards the signs of disease while specific treatment is directed toward the cause of the signs of illness. For example, supportive care for a doe with diarrhea is reducing the concentrate in her diet while, increasing her water intake. But how does one determine if the diarrhea is caused by parasites, toxins, salmonella or overeating? The art of diagnosis depends on careful observation of the goat, its herd mates, its history and its environment to analyze all of the clues to the cause.

Work with a veterinarian or animal technician to reach the best diagnosis and choose the most appropriate treatment. The health charts in this book can be useful in diagnosing certain problems and diseases by way of the signs that are seen. However, in certain cases, laboratory tests are needed for a definitive diagnosis.

In remote areas or field conditions, treatment can begin with an "educated guess" based on common patterns in the area. Remember several disease conditions can co-exist in one animal or herd, complicating diagnosis and making each disease more serious. Poor nutrition and many parasites, for example, can make a case of bacterial pneumonia more serious and harder to treat.

Taking a Goat's Temperature

A goat's normal body temperature is 38.7°C to 40°C or 101.7°F to 104°F. When taking temperatures, a rectal reading from a non-mercury (digital) thermometer is more accurate and better for the environment. If not available, utilize a mercury thermometer to take rectal reading of the goat's temperature.

- *Digital Thermometer*
  - Wipe the thermometer and press the button on the side once.
  - Insert the small end of the thermometer into the goat's rectum.
  - Wait for two minutes. A digital thermometer will beep when the time is complete.
  - Remove thermometer, wipe clean, read and record.
  - Before storing, wipe with antiseptic (alcohol, iodine or bleach).

FAHRENHEIT	CENTIGRADE	
106°F	41.2°C	Goat is sick—has a fever
104°F	40°C	Normal temperature (upper limit)
101.5°F	38.6°C	Normal temperature (lower limit)
98°F	36.5°C	Goat is sick

- *Mercury Thermometer*
  - Wipe or wash the thermometer to clean before use.
  - Shake the thermometer down to below normal, with a quick snap of the wrist. Wet the small end of the thermometer with petrolatum or water.
  - Insert the small end into the goat's rectum.
  - Wait two minutes. (Hold the thermometer or watch carefully to make sure it does not fall out and break.)
  - Remove thermometer, wipe clean, read and record.
  - Caution! Thermometer breaks easily.

USE OF MEDICATIONS, ANTIBIOTICS AND DEWORMERS

Always store all antibiotics, dewormers and other medications out of reach of children.

Before administering any medication, consult an animal health worker or veterinarian. Completely read all instructions accompanying medicines or vaccines. Careful attention should be given to the proper dosage for each animal being treated. Dosage is generally based on the age and weight of an animal and will be detailed in the instructions.

Another important consideration is the withdrawal time, or how long the goat's milk and meat remain contaminated after the drug has been administered. A medicine may remain in the animal's body for as little as a few hours or as long as several weeks, depending on its particular characteristics. It is important not to use milk or meat from a goat before the full withdrawal time. Meat that still contains penicillin and ampicillin, for example, can cause severe allergy in humans. Chloramphenicol can cause severe anemia in humans.

A veterinarian or extension worker can also give advice on withdrawal times if the information is not clearly detailed in the medicine's instructions. If an animal health worker is not available when treating with antibiotics and anthelmintics (dewormers), check the packaging for instructions on withdrawal time. If the withdrawal time is not listed in the instructions, follow these general rules:

- milk withdrawal – three days
- meat withdrawal – two weeks (4 weeks if using streptomycin or thiabendazole; five weeks for ivermectin.)

HERBAL REMEDIES

In each part of the world, plants are used to prevent or treat disease in animals. These are called herbal remedies. An example is the plant aloe vera, found in many warmer climates, which is an excellent healer of wounds and burns. Many such plants are used in the manufacture of commercial medicines. Local farmers, as well as healers and animal health workers may know useful plants in the surrounding area. Some are better researched than others. The mechanism of action of some herbal remedies has been studied, as seen in the table below:

Antimicrobial Plant Components and Their Suspected Mode of Action

COMPONENT	MODE OF ANTIMICROBIAL ACTION	CONTAINED IN
Caffeic acid	Oxidation	Tarragon, thyme
Flavanoids	Plant response to microbial infection	Green tea
Condensed tannins	Bind bacterial cell walls	Sericea lespedeza, oak leaves
Coumarins	Stimulates macrophages	Cured hay (responsible for odor)
Terpenes	Microbial cell membrane disruption	Basil, chili peppers, wormwood (artemesia)
Berberine	Antibacterial/antiamoebic	Goldenseal, barberry, Oregon grape
Lectins and polypeptides	Prevent adhesion of microbes to host receptors	Amaranth, barley, wheat

SOURCE: ANN WELLS, DVM, SPRINGPOND HOLISTIC ANIMAL HEALTH

HOW TO DRENCH AN ANIMAL

Liquid medications, like dewormers, can be given as a drench. Use a dose syringe, a soda bottle or other type of long-necked bottle.

Measure the correct amount of drench into the syringe or bottle. It is very important to keep goat's head in normal position. If head is tipped up, the drench could go in the lungs and cause pneumonia. Place end of syringe or bottle in left side of mouth on back of tongue. Slowly pour drench into the back of the mouth and watch on the left side to be sure the goat is swallowing. Stop temporarily if the goat coughs or chokes.

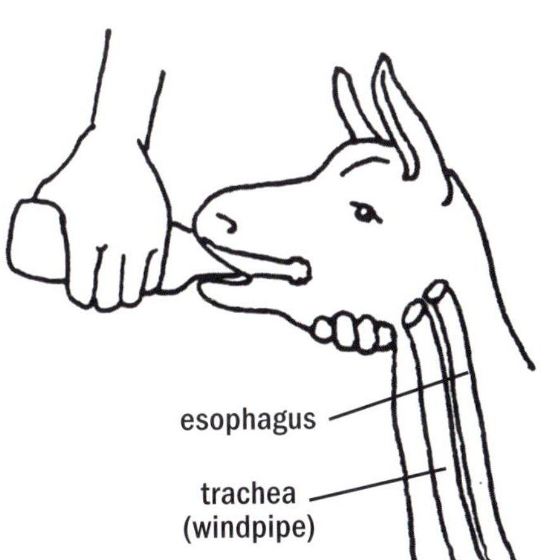

HOW TO PASS A STOMACH TUBE

A stomach tube can be used to remove gas in case of bloat. To do so, use a clean ½ to ¾ inch diameter rubber tube with smooth ends. Measure the length of tube needed to reach the rumen against the outside of the goat and mark the tube with tape. Have someone firmly restrain the goat. Use a speculum or a piece of bamboo through which to pass the tube in order to prevent the goat from biting and damaging the tube. Slowly pass the tube past the bulge in the tongue, down the throat, and continue to the rumen. The tube can be felt on the left side of the neck in the esophagus. If the tube cannot be felt, it may be in the trachea, which is in the middle of the neck.

Stop inserting at the tape mark. Place your ear on the left side and have someone blow into the tube—bubbling sounds near the rumen should be audible.

Be certain the tube has not entered the lungs. If the tube has entered the lungs, the goat will cough, air will move in and out of the tube, and no bubbling will be audible when air is blown into the tube. Remove the tube immediately and try again.

When removing the tube from the rumen, fold the top of the tube over on itself to create a vacuum and prevent rumen liquid from coming out, possibly causing pneumonia.

When treating bloat, if gas is not passed remove the tube and check for froth. Mineral oil can be passed into rumen through the tube to treat for this condition.

HOW TO GIVE AN ANIMAL A BOLUS

A bolus is a pre-measured dose of medicine shaped into a solid cake to be swallowed. It is easier to give an animal a bolus with a "balling gun" or a tube made especially for this purpose. A balling gun can be made by using two pieces of plastic pipe which slide into each other. The size of the tubes will be approximately 1.2 cm and 2.5 cm in diameter.

Remove bolus of the proper dosage from container. Break bolus if needed. Insert bolus in end of balling gun or tube. Hold tube upward to prevent bolus from dropping out. Put tube in left side of goat's mouth and push gently, but firmly past the large base of the tongue. This is important because if the balling gun has not passed the base of the tongue, the goat will eventually spit the bolus out. Push plunger forward forcing bolus over the base of the tongue and into esophagus. Push slowly so as not to injure the goat's throat.

Hold goat's mouth closed and stroke throat downward until bolus is swallowed. Wait to see a clear swallowing motion, to ensure the goat is not holding the bolus in its mouth and will not spit it out. (If the goat spits it out, pick up the bolus and try again.)

HOW TO GIVE AN INJECTION

Medicine, needles and syringes should be stored in a clean, dry cabinet or drawer that is free of dust and dirt. Always use a sterile needle and syringe. When giving multiple injections, use one sterile needle and syringe per animal, as some diseases are transmitted via blood or contaminated needles. Needles and syringes can be sterilized by submerging in boiling water for 20 minutes. They can also be cleaned or sterilized by submerging in alcohol.

There are three main types of injections: intravenous (IV), intramuscular (IM) and subcutaneous (SubQ). Only a veterinarian or technician should administer intravenous, or IV injections because in this technique a needle is used to puncture the animal's vein and medicine immediately enters the blood stream.

Before administering any injection follow the preparations outlined below:

- Collect required materials and medicine in advance
- Read instructions on medicine
- Restrain and weigh the animal (Use weight tape method described in Lesson 1)
- Determine appropriate dosage based on the animal's weight
- Wipe the top of the drug bottle with alcohol
- Pull appropriate dosage of drugs into the syringe using sterile technique
- Select the proper place for injection (give in same location on each goat)
- Wipe the site with alcohol or tincture of iodine

Then administer intramuscular and subcutaneous injections according to the steps outlined below:

IM – Intramuscular

- Use a new or sterilized 20 gauge 1 inch needle for each adult goat
- Use a new or sterilized 20 or 22 gauge ½ to 1 inch needle for each kid
- Attach needle to syringe then insert in muscle in the heavy part of neck or thigh in the same spot on each animal (therefore if a lump appears, the cause will be evident)
- Draw back on plunger to check for blood. If blood flows into syringe, withdraw needle and replace in another site
- Inject drug slowly
- Pull out needle and massage site of injection

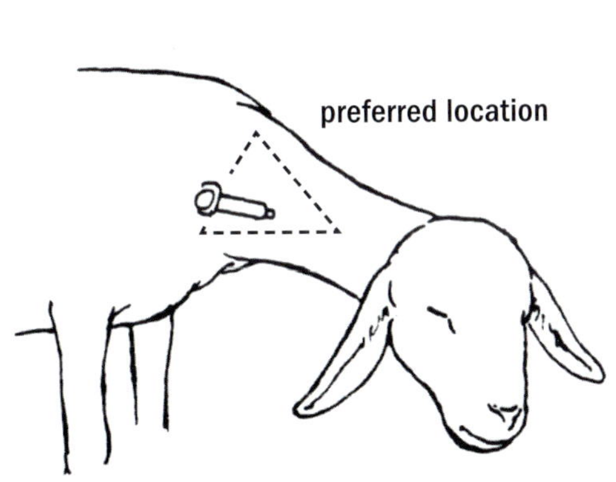

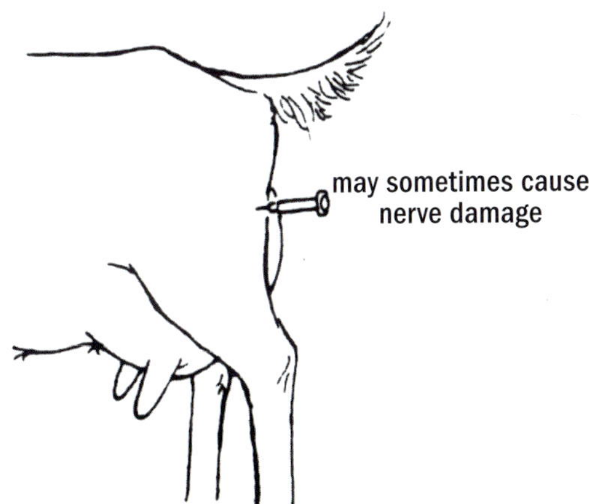

SubQ. – Subcutaneous

- Use the same size needles as for IM injections
- Lift loose skin of the flank, over rib cage, or underneath leg
- Insert needle at acute angle
- Draw back on plunger to check for blood as with IM
- If bubbles appear, the needle may have passed through skin to outside
- If performed properly, a bulge should appear under the skin as the plunger is depressed

PREVENTION AND TREATMENT OF INTERNAL PARASITES

Parasites are small animals that live inside or outside a goat, and get all of their nutrition from the goat's body. They cause disease by consuming nutrients that the goat needs, or by causing intestinal blockage or diarrhea. External parasites like ticks and flies can also transmit infectious diseases.

Both internal and external parasites can cause weight loss and poor growth by consuming a goat's nutrients. Therefore, prevention and treatment is important to the productivity of the herd.

Internal parasites are found in the stomach, intestines or lungs. These include roundworms, lungworms, tapeworms, coccidia and flukes. Certain internal parasites, especially the roundworm *Haemonchus contortus* suck blood, causing severe anemia in goats, which can lead to death. Others like Ostertagia and Trichostongylus cause diarrhea, poor growth and a milder form of anemia. Liver flukes reside in the liver and gall bladder of goats, causing pain, loss of appetite, poor production and sometimes death.

The best prevention for any type of parasite is good management such as nutrition, sanitation, pasture rotation, zero grazing, keeping feed off of the ground and regular fecal exams for parasite eggs. Other important tips to prevent internal parasites are:

- Practice pasture rotation
- Two days after deworming, turn goats out into clean pasture
- Zero grazing pens with slatted floors, and cut-and-carry feeding
- Adequate and balanced nutrition for your goats
- Alternate grazing and cropping of land
- Isolate any animals that have weight loss, diarrhea or anemia
- Give supportive care and consult a technician to determine the cause
- Deworm goats at the appearance of any signs of internal parasites such as anemia, weight loss, poor growth, rough hair coat or bottle jaw. Tropical and other wet climates require more frequent deworming than in cool, dry climates
- Use the FAMACHA* chart to evaluate anemia, if available

* *FAMACHA—A testing system which utilizes an eyelid membrane color chart to determine the level of anemia in a goat. The color of the inside of the eyelid helps to determine the damage caused by the blood-sucking roundworm* Haemonchus

contortus (also referred to as wireworm or barber pole worm), which is the most prevalent and most damaging roundworm in goats. The presence of other roundworms cannot be determined with the FAMACHA Chart. The kit can be obtained through veterinarians and other health practitioners, as it requires training for correct use. Only an original chart can be used; photocopies can give an inaccurate diagnosis. To order the chart, e-mail famacha@vet.uga.edu

If internal parasites are still a problem in the area, treatment can be administered according to the chart below.

Treatment Chart for Internal Parasites

GOAT DEWORMER PERIOD BEFORE SLAUGHTER	TYPE OF INTERNAL PARASITE	SAFE IN PREGNANT ANIMALS
Albendazole (10 days)	■ Roundworms (Nematodes) ■ Tapeworms ■ Liver Flukes	Not in first two months of pregnancy
Fenbendazole (8 days)	■ Roundworms ■ Certain Tapeworms ■ Lungworms	Yes
Levamisole (3 days)	■ Roundworms ■ Lungworms	Yes
Oxfendazole (4 weeks)	■ Roundworms ■ Tapeworms ■ Lungworms	Yes
Morantel Tartrate (30 days)	■ Roundworms	Yes
Ivermectin (5 weeks) Doramectin	■ Roundworms ■ Lungworms (Dictyocaulus type) ■ Mites (some types)	Yes

There are certain plants that are known to have anthelmintic properties. Some of these are *Dorycnium pentaphyllum* (prostrate canary clover); *Onobrychis vicifolia* (Sainfoin), *Lotis pedunculatus* (giant trefoil, maku), *Rumex* sp. (Dock), *Dorycnium rectum*, *Lotus corniculatus* (birdsfoot treroil), *Cichorium intybus* (chicory), *Lespedeza cuneata* (Sericea lespedeza) and neem.

Parasite resistance to dewormers is now a worldwide problem. Frequent deworming encourages resistant worms, which pass on their survival ability to the next generation of worms. Goats with resistant worms will continue to look unthrifty, anemic and thin one week after deworming. You need to deworm these goats again with a different dewormer (not the whole herd). Put these newly dewormed goats back with their herdmates on the original pastures.

Note: In the past, experts recommended changing to a different dewormer each year. The new recommendation is to use the same dewormer until it is no longer effective in your herd. Observation and the use of the FAMACHA chart and fecal egg counts will help determine this.

The following illustrations show how stomach worms and liver flukes feed on your animal.

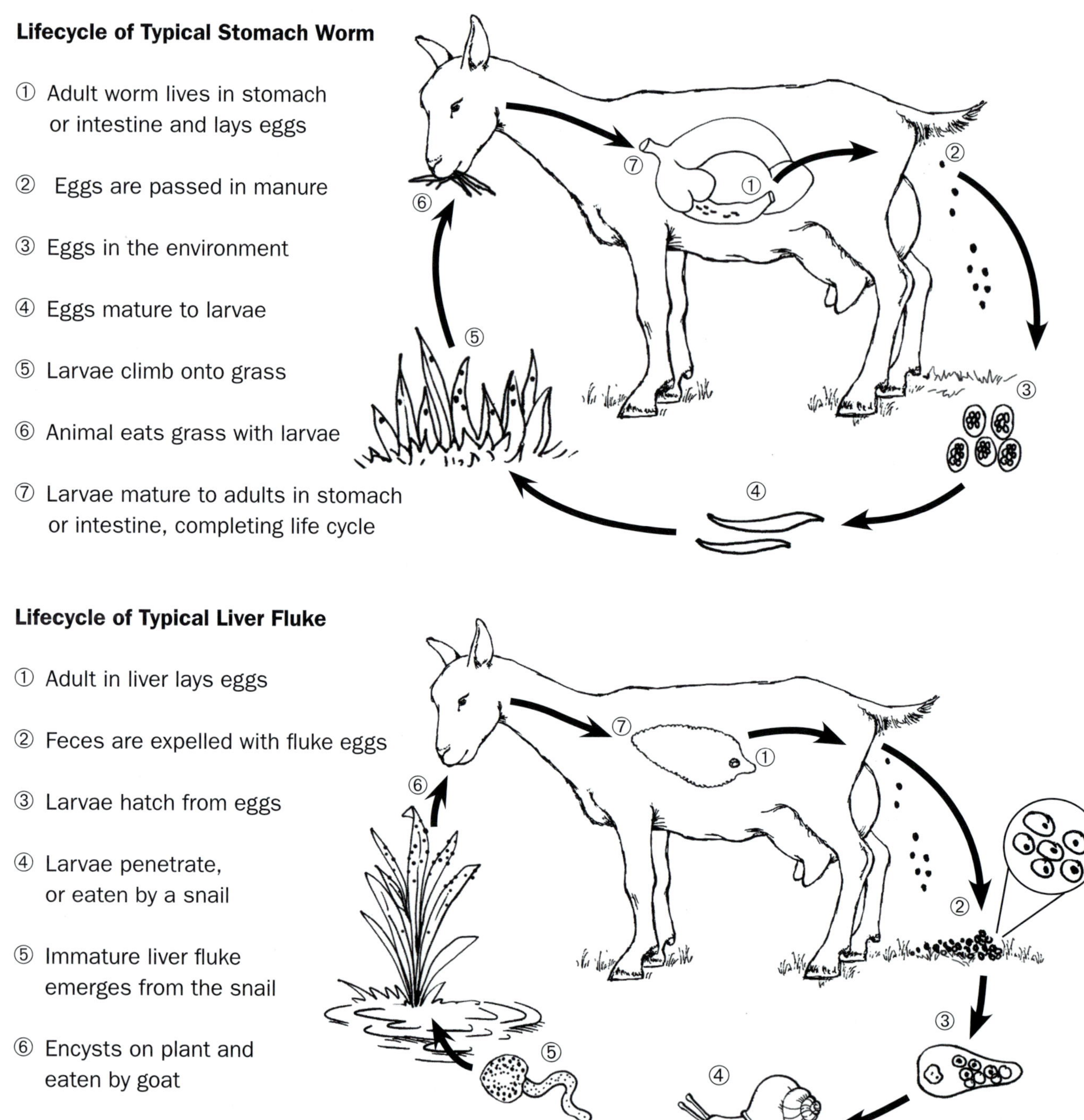

PREVENTION AND TREATMENT OF EXTERNAL PARASITES

External parasites are ticks, lice, mites, fleas and flies. Ticks are a serious problem, particularly in the tropics. Ticks attach themselves to the goat, feeding on its blood, robbing nutrition and transmitting fatal diseases. These diseases include anaplasmosis, babesia, heartwater, and tick-borne fever. Other external parasites that can spread disease between goats during their feeding activities are mosquitoes and tsetse flies.

Some common management strategies for reducing ticks include pasture rotation or zero grazing and managing free range poultry on the same fields as your goats.

Dipping and spraying with insecticides are also commonly used to control tick outbreaks. Depending on the level of infestation, it may be necessary to dip or spray as frequently as every two weeks. In areas where ticks and other external parasites are a problem, they are often found at different locations on an animal's body, including the ears, face and nostrils. As insecticides are poisonous and can cause severe illness to people, special care in storing, handling and disposing of them is recommended. Consult a local veterinarian for advice.

Lifecycle of a One-host Tick

① Eggs hatch. Larvae emerge and climb on goat to feed on animal's blood

② Engorged larvae moult, nymphs emerge and feed on blood

③ Engorged nymphs moult and adults emerge

④ Adults feed on blood and mate

⑤ Engorged female falls to ground and lays eggs and cycle begins again

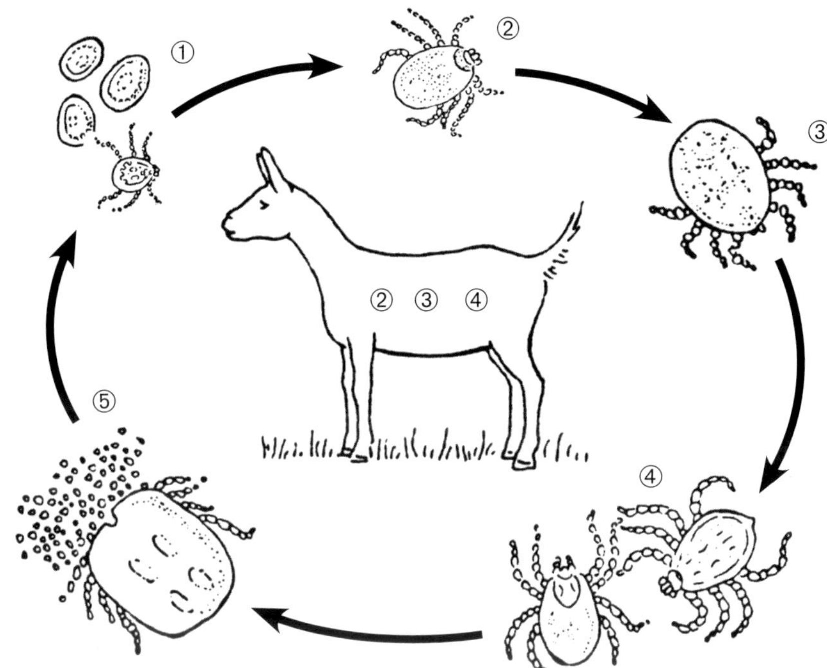

Precautions in Handling Insecticides

When using dip and spray compounds, be sure to follow label recommendations. Many insecticides that kill external parasites are also toxic to other animals, fish, birds and often people. Store all insecticides out of reach of children. Clearly label any containers with the word "poison" to prevent inappropriate usage.

When mixing or spraying, farmers should protect their skin with gloves and other protective clothing. When spraying, wear gloves and a mask to avoid the mist and other exposure. After handling insecticides, bathe thoroughly with soap and water after completion. Mixing equipment should be washed thoroughly and waste water should be buried.

Used product containers should be burned. Using them again is hazardous. Discard any excess dip or spray or other insecticide away from water, gardens or areas where children play. A "hazardous waste disposal site" in the town or village is the safest method of preventing environmental contamination.

Milk withholding times for most insecticides on goats have not been scientifically determined, thus use of any product on goats should be accompanied by an extended period before milk is used for human consumption. The table below lists the most common insecticides in use. There are many more, but please note that Lindane and Methoxychor are not recommended due to possible environmental contamination. Be sure to follow all directions closely to optimize the effect of the insecticide.

Selected Insecticides for Treating Goats with External Parasites

INSECTICIDE	FORMULATION	GENERAL SAFETY
Coumaphos	Powder or liquid	Dilute according to instructions.
Lime Sulfur	Liquid	Occasionally irritating. Otherwise very safe.
Malathion	Powder or liquid	One of the safer Organophosphates.
Pyrethrins and Pyrethroids (Deltamethrin)	Liquid	Pyrethroids cause less skin irritation than Pyrethrins.
Trichlorfon	Powder or liquid	Safe when used as directed.
Neem	Solution in water	Very safe. Requires three treatments. Not effective against mites.

Herbal Insecticides

There are also local treatments that have shown some efficacy for external parasites that may be less toxic and less expensive. One example is the oil from citrus peels. Another is the leaves and seed kernels from the neem tree, listed in the table above. The leaves can be boiled and the resulting tea sprayed on goats to control lice. There may be other herbal treatments used in your area. Below are some herbal remedies for controlling flies.

HERBAL REMEDIES FOR FLY CONTROL		
Cedar oil	Spray or wipe on as needed	Repellant
Garlic and water slurry	Spray or wipe on as needed	Repellant
Citronella, eucalyptus, lemongrass essential oils in mineral oil, or water with soap	Spray or wipe on as needed	Repellant
Neem oil and water	Spray or wipe on as needed	Repellant

SOURCE: ANN WELLS, DVM, SPRINGPOND HOLISTIC ANIMAL HEALTH

OBSERVATION SHEET

Name of Observer: ______________________ Observation Date: __________

Goat's Name: ______________________ Tag #: __________

Breed: ______________________ Age: __________

Tattoo: ______________________ Sex: ❑ Male ❑ Female

Weight: ______________________ Female: ❑ Bred ❑ Open

Last Parasite Treatment: Internal ____/____/____ External ____/____/____

Feeding Plan: ______________________

Observe your goats each day. Record the data at least twice a month. More often if a goat is ill, not thrifty or off feed. Train yourself to evaluate your animals. It is an important step toward herd health and productivity.

General Appearance: ______________________

____

Activity Level: ______________________

____

Legs and Hooves: ______________________

Any external problems such as abscesses, parasites, etc.: ______________________

____

Condition of eyes, nose, and mouth: ______________________

Condition of feces (droppings): ❑ pellet-like ❑ ploppy ❑ loose ❑ diarrhea ❑ __________

Milk Production – if lactating: ________ pounds ________ kilos or ________ liters per day

Any signs of mastitis: ❑ Yes ❑ No

Eating habits: ❑ Eats well ❑ Water available ❑ Range fed ❑ Zero grazing

Check any of the following signs of illness:

❑ Off feed

❑ Dehydration

❑ Limping

❑ Blindness

❑ Diarrhea

❑ Runny eyes

❑ Circling movements

❑ Clots or bloody milk

❑ Standing off from group

❑ Abnormal temperature __________

❑ Pale mucosa around eyes and in mouth

❑ Heavy mucous in nose or mouth

❑ No sign of cud chewing

❑ Swelling at any point of body

❑ Hair falling out or rough in appearance

Refer to the health charts for diagnosis and suggested treatment.

A GUIDE TO DISEASE DIAGNOSIS AND TREATMENT

The following information describes how to use this health care guide for goats. However, common health problems and treatments may differ by region. Always consult local animal health professionals.

You will find many of the diseases, common causes, signs, prevention and treatment listed more than once as they reflect several clinical signs which you can observe. These duplicate listings are for your convenience.

HEALTH PROBLEM	COMMON CAUSES AND CHARACTERISTICS	OTHER SIGNS	PREVENTION	TREATMENT
In this column a common clinical sign of disease, or a common problem, will be listed in capital letters such as: ■ Abnormal behavior ■ Abortion ■ Anemia ■ Cough or nasal discharge ■ Diarrhea ■ Dystocia ■ Eye discharge ■ Fever and poor appetite ■ Infertility ■ Lameness ■ Unwilling to stand ■ Dystocia ■ Skin problems ■ Sudden death ■ Swelling of belly ■ Swelling of udder ■ Weight loss	In this column the diseases or conditions that may cause the sign or health problem in column one will be noted. Remember that in your area there may be other causes of health problems that are not listed here.	This column will list most of the signs of the disease. Your goat may not have all the signs, but only some of them. Observe your animal very carefully to determine all of the visible signs of the disease. Then use this column to help you make the correct diagnosis.	Here are the suggestions for good management which will help prevent the disease. Remember it is always less expensive and more effective to prevent a health problem or disease than to treat one.	There may be alternative treatments for each of these conditions. While it is always good to consult an animal health worker or veterinarian, this may not be possible in remote areas. Local remedies and local knowledgeable healers can be of great assistance.

HEALTH PROBLEM	COMMON CAUSES AND CHARACTERISTICS	OTHER SIGNS	PREVENTION	TREATMENT
Abnormal behavior	**Pregnancy toxemia, (Ketosis)** ■ Can occur last 6 weeks of pregnancy ■ Most common within 10 days of kidding	■ In third trimester ■ Sweet smelling breath ■ Off feed ■ Lack of coordination and other neurological problems such as tremors or stargazing ■ Normal temperature	■ Prevent obesity in early pregnancy ■ Daily exercise ■ Increase plane of nutrition in last two months of pregnancy	■ C-section if caught early enough
	Listeriosis (Circling disease)	See "Fever" section		
	Rabies ■ Fatal to humans	■ Change of behavior and/or aggressive attacks ■ Progressive paralysis ■ Excess salivation ■ Progresses to death within 1-5 days	■ Annual vaccination in areas where rabies is a problem	■ NOT TREATABLE ■ Animals must be slaughtered. ■ Check with doctor when there is human contact with affected animals
	Urinary Calculi ■ In bucks or castrated males	■ Animal distressed and tries to urinate ■ Restless, straining, licking belly ■ Urine may dribble, be bloody and leave crystals in belly hair ■ Death occurs 5-7 days	■ Provide plenty of clean water and salt block or ■ Add common salt to ration ■ Exercise ■ Decrease concentrate. ■ Watch Ca to P ratio – should be 2 to 1	■ If calculus in urethral process, snip it off (used to salvage animals for slaughter) ■ Other surgery expensive, usually not effective long term
	Polioencephalo-malacia ■ Seen in 2- to 6-month old kids fed concentrate	■ Diarrhea ■ Blindness, head pressing ■ Convulsions, coma and death	■ Avoid sudden changes to high energy concentrates ■ Thiamine supplementation ■ Avoid amprolium	■ Thiamine injection IV followed by IM or SubQ every 6 hours ■ Supportive care

HEALTH PROBLEM	COMMON CAUSES AND CHARACTERISTICS	OTHER SIGNS	PREVENTION	TREATMENT
Abnormal behavior (Continued)	**Gid (coenurosis)** ■ Rare in North America ■ From tapeworm (taenia) in dogs	■ Abnormal behavior for a month or more ■ Head tilt, circling, blindness in one eye, unable to stand, rigidity	■ Keep away from stray dogs ■ Treat dogs that live near livestock for tapeworms	■ No effective treatment unless the cyst is visible on head and skull is softened - then aspiration of cyst is sometimes possible
Abortion Abortion may occur naturally in about 15 percent of pregnancies It is very difficult to determine the cause of abortion without specific veterinary tests	**Anaplasmosis** ■ Carried by ticks and possibly other insects and by blood contaminated instruments and needles	■ High fever ■ Pale mucosa (anemia) ■ Abortion in acute cases ■ Animals die quickly	■ Tick control by regular dipping and spraying ■ Blood test animals annually and remove infected animals ■ Clean dehorning and other instruments between use on each animal	■ Oxytetracycline or tetracycline-hydrocholoride or imidocarb
	Brucellosis ■ Caused by *Brucella melitensis* ■ Spread in urine, milk and vaginal discharges ■ Contagious to people via unpasteurized milk. Called Malta fever in humans	■ Lameness ■ Late abortions ■ Mastitis in some cases ■ Joint problems ■ Swollen testicles in males	■ Vaccination where permitted ■ Keep animals away from aborted fetus, membranes and uterine discharge ■ Send tissue samples to veterinarian ■ Bury or burn all other infected tissue ■ To prevent human infection boil or pasteurize all milk before consuming ■ Wear gloves when handling aborted fetuses and membranes	■ No effective treatment ■ Cull

HEALTH PROBLEM	COMMON CAUSES AND CHARACTERISTICS	OTHER SIGNS	PREVENTION	TREATMENT
Abortion (continued)	**Chlamydia psittaci** ■ Contagious to people, especially dangerous to pregnant women	■ Fever, poor appetite ■ May have bloody discharge ■ Up to 60 percent may abort, usually third trimester, placenta abnormal	■ If diagnosed, annual vaccination recommended but causes soreness. Give 8 and 4 weeks before breeding. ■ Remove affected does from herd for 3 weeks	■ Treat all unaffected does with tetracycline 4 weeks before end of gestation if chlamydia is diagnosed in herd
	Leptospirosis ■ Contagious to people via urine	■ Anemia, jaundice ■ Drop in milk production. Milk may be thick, yellow or blood-tinged with no signs of mammary inflammation	■ Keep animals away from urine of infected animals (rodents) ■ Vaccination ■ Rodent control	■ Penicillin or dihydro-streptomycin
	Listeriosis (Circling Disease) ■ Etiologic agent is found in soil and silage ■ Spread by urine, milk and vaginal discharges ■ Contagious to people from milk	■ Abortion with no other signs ■ Circling in one direction ■ Fever over 104°F to 107°F (40.2°C to 43.2°C) ■ Stands alone, lack of appetite ■ Stringy salivation ■ Nasal discharge ■ Paralysis of jaw, eye, and ear muscles, usually on one side only ■ Some recover or death occurs in 1-2 days	■ Discontinue feeding of silage ■ Keep feeders clean ■ Assure adequate nutrition so goats do not consume dirt (pica)	■ High doses of penicillin or tetracycline may help, but often are unsuccessful ■ Fluids and oral rehydration solution (see page 143)
	Poor Nutrition	■ Unthrifty ■ Weight loss ■ Stunted growth	■ Balanced nutrition, including source of protein, energy, vitamins and adequate water	■ Improve nutrition of doe. Combine digestible grasses for energy and leguminous plants for protein ■ Treat for parasites

HEALTH PROBLEM	COMMON CAUSES AND CHARACTERISTICS	OTHER SIGNS	PREVENTION	TREATMENT
Abortion (continued)	**Poisons, toxic plants, insecticides or certain drugs**	■ Abortion after contact with substances	■ Avoid use or ingestion of toxic substances ■ Learn poisonous plants in the area and avoid feeding. Ask local animal health staff	■ Read labels to see if antidote exists ■ Check with local animal health professional
	Toxoplasmosis ■ Transmitted from cat feces. Can be passed to human through raw milk or contact with cat feces. Can cause abortion or birth defects in humans	■ Abortion in last two months of pregnancy ■ Other signs are seldom seen ■ Fever ■ Lack of appetite	■ Prevent contact with cat feces	■ Not usually treatable
	Trauma (injury)	■ Overcrowding, aggression, can cause injury to pregnant animals	■ Avoid overcrowding, ■ Keep pregnant does apart	■ Treat injuries, but once aborted, fetus is lost
Anemia or weakness	**Anaplasmosis** ■ Carried by ticks, (and perhaps other insects) and blood-contaminated instruments and needles	■ Conjunctiva and gums pale or occasionally yellow (jaundice) ■ High fever ■ Abortion in acute cases ■ Animal may die quickly ■ More severe in adult animals	■ Control of ticks and flies by regular dipping or spraying with insecticide ■ Clean dehorning and other instruments between use on each animal	■ Oxytetracycline or tetracycline-hydrocholoride or imidocarb ■ Good nursing and supportive care
	Babesiosis ■ Carried by ticks and contaminated needles and instruments ■ Common in the Middle East, parts of Africa and Indian subcontinent	■ More severe in animals 6-12 months old ■ Pale mucosa (gums and conjunctiva) or jaundiced (yellow) ■ Red-brown urine ■ High fever ■ Lack of appetite ■ Diarrhea ■ Abortion ■ Collapse, coma and death	■ Reduction of Boophilus ticks ■ Regular dipping or spraying	■ Difficult to treat ■ Diminazene, acaprin, or pirevan or imidocarb very long withdrawal time before slaughter—check directions on label

HEALTH PROBLEM	COMMON CAUSES AND CHARACTERISTICS	OTHER SIGNS	PREVENTION	TREATMENT
Anemia or weakness (continued)	**Coccidiosis** ■ A common cause of diarrhea in kids (a parasite of the intestines)	■ Seen in young kids ■ Watery diarrhea ■ Bloody or mucous in feces ■ Weak ■ Unthrifty ■ Weight loss ■ Lack of appetite ■ Dehydration ■ Coccidia seen with microscope in fecal exam	■ Good sanitation ■ Do not overcrowd ■ Keep pens dry and clean ■ Avoid contamination of feed and water with manure ■ Separate kids from older animals ■ Isolate sick animals ■ Provide clean water	■ Amprolium (do not use too long) or decoxquinate or sulfadimethoxine or certain other sulfa drugs or monensin ■ Stop grain and alfalfa feeding while problem exists ■ Give oral rehydration solution free choice or by bottle 6 times daily (see page 143)
	Lice, mites and other external parasites	■ Animal picks at hair ■ Pale mucosa (gums and inside eyelids) ■ Animal rubs against fences, trees or feeders ■ May cause death, especially young animals	■ Good nutrition and sanitation ■ Routine treatment with powders, dip or spray	■ When necessary, spray, powder, or dip animals with local or commercial insecticides ■ Follow all warnings and precautions
	Poor Nutrition ■ Especially iron deficiency	■ Signs similar to parasites	■ Assure balanced feed in sufficient quantity ■ Iron is found in green leaves and tomatoes	■ Assure balanced feed in sufficient quantity
	Stomach worms **Intestinal worms** **Liver flukes**	■ Pale mucosa ■ Weight loss ■ Swollen belly and below chin ("bottle jaw") ■ Rough hair coat ■ Unthrifty ■ Poor growth in kids ■ Diarrhea ■ Reduced appetite	■ Parasite control program ■ Good nutrition ■ Rotate and manage pastures or tether goats in a new place each day ■ Cull heavily infected animals	■ Fenbendazole ■ Ivermectin or doramectin injections or levamisol (Tramisol) or albendazole *(Note: Not all the drugs effective against worms are effective against flukes – read labels* ■ See treatment of liver flukes under "weight loss" section

HEALTH PROBLEM	COMMON CAUSES AND CHARACTERISTICS	OTHER SIGNS	PREVENTION	TREATMENT
Cough and nasal discharge	**Lungworms**	■ Cough ■ Labored breathing ■ Weight loss ■ Nasal discharge ■ Signs similar to pneumonia	■ Prevent access to wet pasture ■ Deworm regularly	■ Deworming program with levamisol, fenbendazole or oxfendazole
	Pneumonia ■ Caused by bacteria, viruses or mycoplasma.	■ Normal to high fever ■ Cough, nasal discharge ■ Rapid, exaggerated or "labored" breathing ■ Raspy sound in lungs ■ Tearing of eyes ■ Grinding teeth ■ Lack of appetite ■ Lack of energy ■ Unable to stand	■ Draft-free housing ■ Good ventilation ■ Protect from wind and rain ■ Do not overcrowd ■ Reduce stress ■ Mycoplasma vaccine specific for CCPP organism	■ Antibiotics such as: tetracycline, tylosin, penicillin, ampicillin or penicillin-streptomycin ■ Sulfonamides (use with kids) ■ Good nutrition ■ Plenty of clean water
	Contagious Caprine Pleuro-pneumonia ■ A type of pneumonia causing death within two days	■ Similar signs to pneumonia but more severe and acute ■ Standing with elbows out ■ Reluctance to move due to pain of pleuritis		■ Tetracycline or tylosin
	Goat pox can also cause cough or pneumonia	See "Skin looks abnormal" section		
Diarrhea	**Coccidiosis** ■ A common cause of diarrhea in kids. A parasite of the intestines	■ Seen in young kids ■ Watery diarrhea ■ Bloody feces or feces with mucous ■ Weak ■ Unthrifty and losing weight ■ Lack of appetite ■ Dehydration ■ Sudden death	■ Sanitation very important ■ Do not overcrowd ■ Keep pens dry and clean ■ Avoid contamination of feed and water with manure ■ Separate kids from other animals ■ Isolate sick animals ■ Provide clean water	■ Give oral rehydration solution free choice (see page 143) ■ Amprolium, decoxquinate, or certain sulfa drugs (sulfaquinoxaline) or monensin ■ Stop grain and alfalfa feeding while problems persist ■ Feed grass hay

HEALTH PROBLEM	COMMON CAUSES AND CHARACTERISTICS	OTHER SIGNS	PREVENTION	TREATMENT
Diarrhea (continued)	**Bacterial Diarrhea such as Salmonellosis (or E. Coli)**	■ Diarrhea bloody (if salmonella) ■ Weight loss ■ Dehydration ■ Fever ■ Loss of appetite ■ Death	■ Isolate sick goats ■ Buy only healthy goats from healthy herds ■ Make sure kids get colostrum at birth ■ Isolate sick goats in clean and dry area ■ Keep all kids clean and dry	■ Antibiotics (tetracycline or neomycin) or sulfa drugs ■ Give oral rehydration solution (see page 143) ■ Offer dry hay and clean water
	Enterotoxaemia (Overeating Disease) ■ Affects animals in good condition	■ Depression ■ Staggering ■ Loss of motor control ■ Stargazing ■ High fever ■ Abdominal pain ■ Foul-smelling diarrhea ■ Death within 24 hours	■ Vaccination—twice two weeks apart and then two, or even three times yearly if severe problem ■ Avoid sudden changes in feed ■ Avoid feeding too much grain	■ No effective treatment
	Poisonous plants ■ Usually more of a problem in dry season	■ Loose stools ■ Bloat, colic ■ Staggering ■ Convulsions ■ Frothing at mouth ■ Pale or blue mucosa (gums and conjunctiva)	■ Identify poisonous plants in the area and keep goats away from them ■ Feed hay before turning goats out to pasture ■ Good feeding practices	■ Give laxatives such as mineral oil ■ Give activated charcoal orally to counteract toxin ■ Treatment varies with plant and has limited effectiveness depending on amount of poisonous plant ingested
	Stomach worms Intestinal worms (nematodes)	■ Anemia especially with *Haemonchus cortortus* ■ Weight loss ■ Unthrifty ■ Poor growth of kids ■ Swollen belly or under jaw ■ Rough hair coat	■ Check fecal sample ■ Use FAMACHA chart ■ Parasite prevention program including pasture rotation, sanitation, isolation of new animals ■ Deworming program if necessary ■ Good nutrition	■ Fenbendazole or ivermectin or levamisol or albendazole or mebendazole

HEALTH PROBLEM	COMMON CAUSES AND CHARACTERISTICS	OTHER SIGNS	PREVENTION	TREATMENT
Diarrhea (continued)	**Sudden changes in feed or intake of lush pasture**	■ Bloat ■ Sudden death ■ History of recent change of feed or pasture	■ Prevent overgrazing ■ Change feed gradually	■ Treat for bloat ■ Turn to dry grass hay
Dystocia (difficult birth) and problems at time of birth	■ Two or more kids in uterus ■ Very large kid(s) ■ Abnormal position of kids in uterus ■ Obesity	■ Labor extends beyond one hour or an abnormal discharge	■ Not easily preventable, so be prepared to assist with birth if necessary to save kids and doe ■ Do not breed young or small doe with large buck ■ Control diet to prevent overweight doe ■ Prevent obesity in early pregnancy	■ Carefully assist the doe in giving birth. See Chapter 6 ■ Caesarian section by veterinarian may be necessary
	Milk Fever Hypocalcemia ■ Rare in goats, occurring mainly in high dairy producers near time of kidding	■ Decreased appetite ■ Mild bloat ■ Constipation ■ Trembling ■ Stiff muscles ■ Staggering ■ Unable to stand ■ Low temperature ■ Increased heart rate ■ Death	■ Low calcium diet in late pregnancy	■ Emergency - treat with slow IV injection of 25 percent calcium borogluconate solution by qualified animal health professional (50-100cc)
	Pregnancy toxemia (Ketosis) ■ Third trimester: Most common within 7 to 10 days of kidding	■ Sweet smelling breath ■ Rejects feed ■ Depression, unable to rise ■ Lacks coordination. Other neurological problems: tremors, stargazing ■ Death	■ Daily exercise ■ Increase plane of nutrition in last two months of pregnancy	■ C-section as soon as possible. ■ 15cc of propylene glycol twice daily ■ Single IV injection of 50 percent dextrose (100-200cc) by trained animal health professional

HEALTH PROBLEM	COMMON CAUSES AND CHARACTERISTICS	OTHER SIGNS	PREVENTION	TREATMENT
Dystocia (difficult birth) and problems at time of birth (continued)	**Retained Placenta (Afterbirth)** ■ Part of birth sac not expelled	■ Part of afterbirth hanging from the vulva, or not found ■ Several days later: vaginal discharge and foul odor. *Note: a bloody discharge called lochia is normal for two weeks after birth, but it does not smell rotten*	■ Good nutrition and general health care	■ An animal health worker may slowly remove the placenta by hand after 5 days ■ Help prevent by adequate selenium and vitamin E ■ Give antibiotics
Eye (ocular) discharge	**Injury to eye**	■ Tears ■ Redness of eye	■ Disbud and dehorn goats to prevent them from harming one another ■ Keep sharp objects out of pens	■ Antibiotic eye ointment such as Terramycin—2 or 3 times daily
	Kerato-conjunctivitis (pink eye) ■ Transmitted by insects ■ Cause may be bacteria, virus, mycoplasma or chlamydia	■ Eyes stuck shut ■ Tearing ■ Red conjunctiva (inside eye lid) ■ Cornea cloudy or opaque ■ Blindness	■ Control flies ■ Separate infected animals ■ Reduce dusty conditions	■ Administer Tetracycline antibiotic eye ointment (e.g. Terramycin) 2-3 times daily. Be sure not to touch eye when applying. If possible, ■ Use separate tube for each infected animal ■ Do not use powders as they irritate eyes ■ Apply patch over eye for 2-3 days
	Pneumonia or other respiratory infections	■ Normal to high fever ■ Discharge from nose, eyes or mouth ■ Rapid, exaggerated breathing ■ Grinding teeth ■ Lack of appetite ■ Lack of energy	■ Draft free housing ■ Protect from wind and rain	■ Tetracycline ■ Tylosin ■ Penicillin ■ Ampicillin ■ Penicillin-streptomycin ■ Sulfanamides (for kids) ■ Good nutrition ■ Plenty of clean water

HEALTH PROBLEM	COMMON CAUSES AND CHARACTERISTICS	OTHER SIGNS	PREVENTION	TREATMENT
Eye (ocular) discharge (continued)	**Hemorrhagic Septicemia**	■ Discharge from eyes ■ Fever ■ Depression ■ Cough	■ Reduce stress ■ Vaccinate ■ Avoid crowding in pens and barns	■ Antibiotics ■ Sulfas ■ Reduce stress
	Peste des Petites Ruminants (PPR)	See description in "Fever" section below		
Fever and poor appetite	**Goats left in hot sun without water**	■ Panting ■ Listless ■ High temperature ■ Sudden death	■ Do not tether goats in sun for long periods without shelter or shade and water	■ Provide shade and clean water
	Any infection including the following: **Anthrax** ■ A worldwide disease in dry areas and after floods—humans can become infected and very sick ■ Do not eat meat of animals which may have died of anthrax	■ Fever ■ Bloody discharge from nose, mouth or rectum ■ Purple gums ■ Convulsions ■ High fever ■ Sudden death ■ Fever—over 104°F to 107°F	■ Isolate sick animals, vaccinate all others ■ Vaccination ■ Isolation and quarantine of infected animals ■ Disposal of infected carcasses by burning or deep burial ■ Do not cut open dead carcasses	■ Oxytetracycline or penicillin at double the normal dose will occasionally save an animal
	Listeriosis (circling disease) ■ Etiologic agent is in soil and silage ■ Spread in urine, milk and vaginal discharges	■ Circles in one direction, head pressing ■ Stands alone, Lack of appetite ■ Abortion with no signs ■ Stringy salivation ■ Nasal discharge ■ Paralysis of jaw, eye and ear muscles on one side only ■ Death in 1-2 days	■ Discontinue silage feeding ■ Keep feeders clean ■ Assure adequate nutrition so goats do not consume dirt (pica)	■ High doses of tetracycline or penicillin may help but often unsuccessful ■ Fluids and oral rehydration solution (see page 143)

HEALTH PROBLEM	COMMON CAUSES AND CHARACTERISTICS	OTHER SIGNS	PREVENTION	TREATMENT
Fever and poor appetite (continued)	**Navel Ill (Joint Ill)**	■ Inflamed, swollen umbilicus, maggots possible ■ Fever in newborns ■ Swollen joints ■ Depression, will not stand	■ Clean, dry environment ■ Dip navel in 7 percent tincture of iodine immediately after birth ■ Repeat in 12 hours ■ Give colostrum as soon as possible after birth	■ Penicillin and streptomycin ■ Supportive care
	Peste des petits Ruminants (PPR) (Pest of small Ruminants)	■ Fever 104°F-107°F ■ Eye and nasal discharge ■ Erosions and ulcers in mouth ■ Pneumonia ■ Diarrhea, abdominal pain ■ Emaciation. Death in 5-7 days or longer	■ PPR vaccine very effective. However, do not vaccinate during outbreak. ■ Do not use Rinderpest vaccine ■ Isolate infected animals. Notify authorities if you suspect this disease	■ No treatment exists ■ Supportive care, improved nutrition, and antibiotics may reduce deaths
	Pneumonia ■ Most severe in young animals	■ Normal to high fever ■ Tearing of eyes ■ Nasal discharge ■ Rapid exaggerated breathing. Raspy sounds in lungs	■ Good ventilation ■ Reduce stress ■ Protect from wind, rain, avoid overcrowding	■ Tetracycline or tylosin or penicillin, or ampicillin, or penicillin-streptomycin ■ Improve nutrition ■ Plenty of clean water ■ Ventilation of barn
	Other causes of high fever: Anaplasmosis, Mastitis, Tetanus, Foot and Mouth Disease, Contagious Caprine Pleuropneumonia, Goat Pox	See other pages in the "Health" section on these diseases		
Infertility, sterility	**Obesity**	■ Excess nutrition during pregnancy or in the breeding buck	■ Proper diet and exercise	■ Limit feed intake until weight loss occurs, but give balanced diet, exercise

HEALTH PROBLEM	COMMON CAUSES AND CHARACTERISTICS	OTHER SIGNS	PREVENTION	TREATMENT
Infertility, sterility (continued)	**Intersex (Hermaphrodites)** ■ An intersex is an animal that shows both male and female characteristics	■ Ovarian and testicular tissue both present	■ Do not breed two polled animals (born without horns) ■ One goat in each breeding pair must have been born with horns	
	Metritis (Infected uterus)	■ Discharge from vulva with blood or pus ■ Strong odor ■ Fever ■ Doe does not come into heat	■ Good sanitation at birth ■ Clean pens ■ Disinfect hands before assisting with birth	■ Antibiotics applied to uterus and by injection ■ Same as prevention
	Poor Nutrition ■ Especially phosphorus deficiency or vitamin E-selenium deficiency	■ Unthrifty ■ Poor growth	■ Balance protein, energy, vitamins and minerals in diet	■ Feed balanced ration
	Testicles not developed	■ No libido ■ Testicles small or absent	■ Avoid in-breeding	■ None
	Illness	■ Signs vary depending on type of illness	■ Use buck in good health ■ All preventive health care will keep your breeding animals healthy and reduce infertility	■ Treat any existing infections before trying to breed animals. It takes 3 weeks after a fever for a buck to regain his fertility ■ Treat any illness as soon as it occurs
	Heat stress	■ Buck does not breed ■ Rapid breathing	■ Keep bucks in shady well-ventilated area ■ Do not breed animals when outside temperatures are excessive	
Lameness	**Arthritis**	■ Swollen joints ■ Pain ■ Difficulty standing and walking ■ Swollen naval	■ Choose sound animals with good bones ■ Do not use concrete floors ■ Trim feet regularly ■ Good nutrition ■ Disinfect navels at birth	■ Trim feet ■ Remove from concrete or other hard surfaces ■ Antibiotics and anti-inflammatory medication (but not steroids)

HEALTH PROBLEM	COMMON CAUSES AND CHARACTERISTICS	OTHER SIGNS	PREVENTION	TREATMENT
Lameness (continued)	**Caprine Arthritis Encephalitis (CAE)** ■ A viral disease caused by a retrovirus most likely transmitted through colostrum and milk	**In kids:** ■ Progressive paralysis starting with weakness in rear legs ■ Lack or coordination ■ Seizures, death ■ No fever **In adult goats:** ■ Swollen, painful joints ■ Goat walks on carpi (front knees) ■ Weight loss, hard udder, little milk	■ Do not let kids suckle ■ Separate kids from doe at birth ■ Feed heat-treated colostrum to kids ■ Feed pasteurized milk or milk replacer to kids ■ To heat treat colostrum: see Chapter 3 ■ If possible, test goats annually ■ Cull animals that are positive to the test ■ Cull animals with clinical signs of CAE	■ No effective treatment
Unwilling to stand	**Hoof rot**	■ Lameness ■ Foul smelling odor from hoof ■ Holding leg up ■ Unable to walk ■ Decrease in production	■ Proper and regular hoof trimming ■ Keep off of wet pastures	■ Trim hooves, ■ Apply antibiotic dressings ■ Put in pen with dry bedding
	Overgrown Hooves ■ Especially a problem in confined goats and goats that do not exercise or graze	■ Lameness ■ Long curled toes "slipper foot"	■ Regular trimming ■ Exercise ■ Flat, smooth wooden slatted floors in pen, not unfinished logs or rounded bamboo	■ Trim feet regularly
	Enterotoxaemia (Overeating disease)	■ See section on Sudden Death	■ Vaccinate	■ No effective treatment
	Milk fever (hypocalcemia) ■ Usually high producing dairy goats	■ Decreased appetite ■ Mild bloat ■ Constipation ■ Trembling ■ Stiff muscles ■ Staggering ■ Unable to stand low temperature ■ Increased heart rate ■ Death	■ Low calcium diet in late pregnancy may help prevent.	■ Emergency - treat with slow IV injection of 25 percent calcium borogluconate solution by qualified animal health professional (50-100cc)

HEALTH PROBLEM	COMMON CAUSES AND CHARACTERISTICS	OTHER SIGNS	PREVENTION	TREATMENT
Unwilling to stand (continued)	**Pregnancy Toxemia**	See section on "Sudden death"		
	Severe Hoof Rot	See section on "Lameness"		
	Advanced CAE	See "Lameness" section for description of CAE		
	Tetanus	■ Lockjaw, rigid legs ■ Stiff rigid stance (legs straddled, neck and head extended) ■ Tetanic contractions of muscle, stiffness ■ Protrusion of third eyelid ■ Fever ■ Sudden death	■ Vaccinate with Tetanus Toxoid–2 doses, 30 days apart	■ Give tetanus anti-toxin, and double doses of penicillin daily for 5 days ■ Keep in quiet place away from sunlight to reduce tetanic muscle contractions ■ Nursing care—offer food, water by hand ■ May need to treat for bloat (stomach tube)
Skin looks abnormal	**Abscesses (Caseous lymphadenitis)**	■ Swollen lymph nodes common under jaw and ear, in front of shoulder, on flank, above udder or scrotum ■ Nodes may be warm or swollen and may contain greenish or cheesy pus	■ Isolate infected animals ■ Avoid contamination ■ Do not buy goats with abscesses or swollen lymph nodes	■ Difficult to treat ■ Surgically open abscesses, irrigate with 7 percent iodine or hydrogen peroxide, flush daily. ■ Separate from healthy goats before opening abscess to avoid contamination. Keep animal separate until healed ■ Burn or bury all material from abscesses
	Contagious Ecthyma (Soremouth) ■ Contagious to people, especially from the vaccine	■ Pustules, and then crusts on mouth, feet, eyelids, and teats ■ Malnutrition ■ Weight loss Dehydration in kids	■ In infected herds, animal health workers should vaccinate all animals first, and then new additions and kids ■ Caution on handling vaccine	■ Supportive care, including hand feeding of kids, until full recovery ■ No other treatment exists

HEALTH PROBLEM	COMMON CAUSES AND CHARACTERISTICS	OTHER SIGNS	PREVENTION	TREATMENT
Skin looks abnormal (continued)	**Ear mites**	■ Shaking head ■ Black crumbly discharge in ear	■ Clean ears once a week with mineral oil	■ Mix cresol with mineral or palm oil and rub in ear
	Goat Pox	■ Circular, raised areas on skin, especially in muzzle and ears ■ Swollen eyelids ■ Pneumonia, discharge from nose ■ May cause death	■ Vaccination with live-virus vaccine ■ Isolate infected goats	■ Spray or dip with insecticides such as coumaphos (Asuntol)
	Lice	■ Animals rub against objects and pick at hair ■ Rough hair coat, patchy loss of hair ■ Lice visible–check skin on back ■ Pale mucosa ■ Can cause death in heavy infestations if blood sucking lice	■ Routine treatment with insecticide, powders, sprays or dips ■ Good nutrition ■ Avoid winter overcrowding of stock	■ Ivermectin injections (observe withdrawal period) ■ Many effective local treatments exist, including preparation made from leaves and seed of the neem tree. Check with experienced farmers about local treatments
	Mange ■ Scab mite (Sarcoptes and Psoroptes) ■ Demodex	**Sarcoptes and Psoroptes:** ■ Severe itching ■ Loss of hair, sometimes extensive ■ Skin dry and flaky ■ Scab formation **Demodex:** ■ Small lumps under skin (Demodex mites)	**Sarcoptes and Psoroptes:** ■ Spray or dip goats regularly ■ Improve nutrition ■ Isolate infected animals ■ Avoid overcrowding of stock (such as in winter)	**Sarcoptes and Psoroptes:** ■ Treat every two weeks with insecticide until mange is gone. Can apply oil mixed with cresol ■ Lime sulfur is very effective for psoroptes ■ Wear gloves when treating ■ Ivermectin for psoroptes and sarcoptes–35 days withdrawal **Demodex:** ■ Demodectic mange is difficult to treat ■ Can open the lumps and apply iodine

HEALTH PROBLEM	COMMON CAUSES AND CHARACTERISTICS	OTHER SIGNS	PREVENTION	TREATMENT
Skin looks abnormal (continued)	**Ringworm** ■ A fungus - disease of skin and hair ■ Contagious to humans and other animals	■ Circular crusty spot often grayish in color and thin hair ■ Itching	■ Keep infected animals isolated, ■ Avoid crowding ■ Dry environment ■ Remove sharp objects from pens	■ Apply iodine, thiabendazole or diluted chlorine bleach—be careful not to get in goat's eyes. ■ Toothpaste containing zinc, may help if applied topically
	Wounds	■ Bloody area on skin ■ May see maggots (fly larvae) ■ Caution – untreated maggots can lead to death ■ Bad smell (if wound infected, or infested by fly larvae)	■ Control flies ■ Remove horns, but not during fly season	■ Wash with soap and water, then disinfectant. Apply protective bandage in fly season. ■ If maggots present, apply screwworm spray, coumaphos, or cresol ■ If fever, use antibiotics ■ If laceration is large, sutures (stitches) may be required to protect the muscle tissue under the skin
Sudden death, after abnormal behavior	**Anthrax** ■ A worldwide disease in dry areas and after floods—humans can become infected and very sick ■ Do not eat meat of animals which may have died of anthrax	■ Fever ■ Bloody discharge from nose, mouth or rectum ■ Purple gums ■ Convulsions ■ High fever ■ Sudden death ■ Fever—over 104°F to 107°F	■ Isolate sick animals, vaccinate all others ■ Vaccination ■ Isolation and quarantine of infected animals ■ Disposal of infected carcasses by burning or deep burial ■ Do not cut open dead carcasses	■ Oxytetracycline or penicillin at double the normal dose will occasionally save an animal
	Bloat ■ Most often caused by lush green pastures or legumes, excess grain or poisonous plants	■ Swelling of paunch (shows on left side) ■ Colic, restless, anxious behavior ■ Difficult breathing followed by death	■ See section below: "Swelling of Belly" for prevention of bloat	■ See section below: "Swelling of Belly" for treatment of bloat

HEALTH PROBLEM	COMMON CAUSES AND CHARACTERISTICS	OTHER SIGNS	PREVENTION	TREATMENT
Sudden death, after abnormal behavior (continued)	**Enterotoxaemia (Overeating disease)**	■ Death within 24 hours ■ Staggering ■ Stargazing ■ High fever ■ Abdominal pain ■ Foul-smelling diarrhea ■ Affects animals in good condition	■ Vaccination ■ Avoid sudden changes in feed ■ Avoid feeding too much grain ■ Give mineral oil after overeating	■ Usually no effective treatment, however an animal health worker may try penicillin, fluids, and supportive, nursing care
Swelling of belly	**Bloat** ■ Most often caused by lush grass pastures or legumes, excess grain or poisonous plants	■ Swelling of paunch (shows on left side) ■ Colic, restless, anxious behavior ■ Difficult breathing followed by death	■ Limit lush pastures and fast growing legumes. Gradually increase time on these pastures ■ Feed non-legume dry hay before putting out to pasture ■ Avoid excess grain ■ Keep away from poisonous plants	■ Emergency – treat rapidly ■ Keep animal on feet and walking ■ Tie stick or rope in goat's mouth forcing goat to salivate ■ Pass a stomach tube. See page 115 ■ Frothy bloat: drench with palm oil, peanut, vegetable oil or mineral oil ■ In emergency (if animal is already down and unable to rise) pass trocar or sharp knife into rumen high on left side and behind last rib and then treat for infection. This should be done by an experienced person
	Stomach and Intestinal worms	■ Diarrhea ■ Anemia ■ Unthrifty ■ Poor growth	See full description in "Weight loss" section	
	Urethral blockage in males	See section on "Abnormal behavior"		

HEALTH PROBLEM	COMMON CAUSES AND CHARACTERISTICS	OTHER SIGNS	PREVENTION	TREATMENT
Swelling of udder	**Mastitis** ■ Inflammation of the mammary gland	■ Udder swollen, hot, red, sometimes hard and painful ■ Fever ■ Abnormal milk (stringy, bloody, watery) ■ Lack of appetite ■ Can cause gangrene or death	■ Sanitation of milking equipment ■ Clean udder (If you choose to wash a goat's teats before milking, use soapy water or iodine with water. Goats are much cleaner than cows and it is usually not necessary to wash the udder. You can brush off any loose dirt and hair. Be sure to dry teats thoroughly. Use a strip cup to examine milk before milking ■ Full hand milking with clean hands. ■ Use teat dip* ■ Milk infected goats last	■ Milk out frequently ■ Massage udder ■ Hot compresses ■ Teat infusion with antibiotics (use ½ cow treatment—most effective during dry period) ■ If fever, systemic antibiotics i.e. penicillin (If culture of bacteria is possible, antibiotics can be more specific) **Teat dip: (Use a commercial dip or use 0.5 percent iodine solution or 0.5 percent chlorhexidine or solution of 5 percent bleach to 95 percent clean water)*
Weight loss	**Internal Parasites (Stomach worms, Intestinal worms)**	■ Anemia—pale mucosa ■ Swollen belly ■ Dull, rough hair coat ■ Diarrhea ■ Unthrifty ■ Poor growth	■ Deworming program ■ Rotate pastures ■ Good nutrition	■ Fenbendazole or ivermectin, doramectin or levamisole or albendazole or mebendazole ■ (Many local treatments exist for parasites)
	Liver flukes ■ Snail is host ■ Flukes cause damage to liver	■ Weakness Depression ■ Abdomen swollen and painful ■ Weight loss, "bottle jaw" ■ Anemia ■ Jaundice ■ May cause sudden death	■ Difficult to prevent. Keep animals out of wet pastures or irrigation ditches ■ Run ducks on wet pastures to eat snails ■ Routine treatments with fluke wormers ■ Good nutrition	■ Albendazole or clorsulon or oxyclozanide or niclofolan or rafoxanide or triclabendazole ■ Fasinex is the best treatment available because it kills all stages of the liver fluke. **Not approved in US**

HEALTH PROBLEM	COMMON CAUSES AND CHARACTERISTICS	OTHER SIGNS	PREVENTION	TREATMENT
Weight loss (continued)	**Poor nutrition** **Overstocking**	■ Signs similar to internal parasites	■ Avoid overstocking pens and barns ■ Assure adequate and balanced nutrition	■ Same as prevention
	Johnes disease (Yo-nees) ■ Paratuberculosis ■ Wasting disease	■ Chronic progressive weight loss ■ Lack of appetite and depression ■ Droppings loose, but diarrhea rare	■ Test and cull infected goats ■ Separate kids from doe at birth. Feed heat treated colostrum and pasteurized milk or milk replacer ■ Vaccine available in some countries (not United States)	■ None
	Lice	See "Skin looks abnormal" section		

ORAL REHYDRATION SOLUTION

Prepare as needed:

- 1 liter boiled water (then cooled)
- 1 tsp salt
- 8 tsp sugar

Mix well.

HEALTH PROBLEM	COMMON CAUSES AND CHARACTERISTICS	OTHER SIGNS	PREVENTION	TREATMENT
Use this page to list other local diseases which are a problem in your area.				

Goats Give Life to a Small New England Farm

Ann Starbard stands in the pasture overlooking Crystal Brook Farm in Sterling, Massachusetts. Goats of all sizes and colors browse the bushes and leaves on the trees and lie in the shade chewing their cud.

A goat jumps up, puts her front feet on Ann's shoulders and looks her in the eye. "For me, goats have a way of reducing stress from the outside world," says Ann. "They seem to understand the simplicity of life and share that with you when you look into their eyes." Ann reflects that she is part of a larger circle of goat farmers that reach around the world.

▲ Ann Starbard (left) and Rosalee Sinn with kid goats.

Ann was introduced to goats while working for Heifer International. With a degree in animal science from Pennsylvania State University she was in charge of acquiring goats and organizing animal shipments for Heifer's Northeast Region in the late 1980's. She visited goat farms, observed many management systems and developed a strong affection and appreciation for goats.

In 1998, Ann and her husband Eric decided to go into full-time farming, beginning with 40 cows, selling the milk wholesale. Feeling the financial pains of a small dairy farm, Ann worked off the farm and helped Eric tend the cows and fields in the morning and evening. Looking for a way to increase profitability for their farm, Ann seized the opportunity to develop a goat cheese business. The 100-acre Crystal Brook Farm provided an ideal setting. So Ann and Eric purchased a herd of 40 goats and cheese making equipment, and embarked on a new life journey. The equipment that came with the goats included a vacuum-bucket milking system, six milking stands, a 150-gallon bulk tank, an 80-gallon pasteurizer and stainless steel buckets. In time, effective management strategies were developed. The goat cheese enterprise proved to be the most profitable component of Crystal Brook Farm so Ann and Eric sold the cow herd. In 2006 the Starbards invested $18,000 to build a modern milking parlor, including a concrete milking platform, an articulated headlock and a pipeline milking system.

Continue on page 146

Continued from page 145

▲ *Goats browsing in pasture.*

The goat cheese business began with 70 percent of the income being derived from sales to a wholesale distributor. Now 75 percent of the income is derived from direct sales at farmers markets. Ann uses only milk from her own herd to produce the goat cheese. This insures the freshness, purity and quality of the cheese product. The concept of eating locally-produced foods has taken a strong hold in the region, opening up additional opportunities in urban farmers' markets.

Ann currently milks 65 dairy goats of mixed breeds—Saanens, Alpines and LaManchas. Each goat produces an average of one gallon of milk daily and each gallon produces one pound of fresh cheese. Ann uses this basic ratio to estimate production costs and profit and to make adjustments to herd husbandry. "Goats' output in terms of milk and reproduction versus their body weight amazes me!" Ann exclaims.

The herd is bred in the fall and all goats are dried off in the winter for seven to eight weeks. Currently, breeding is done with natural service. The does kid in late winter and milk for 10 months.

The herd receives four to five pounds of home-grown grass/legume hay daily. They have free access to seasonal browse using a rotational browsing system. Ann believes the fresh herbaceous material available in the browse area is the best source of nutrition for the goats. "My favorite part of the day is listening to the goats chew bramble leaves, fresh grass, or winter hay. I think of the thousands of people around the world who enjoy this same experience. The goats are magical animals and I feel blessed to have them." In addition, the goats are fed one to three pounds of a feed concentrate per day and receive free choice salt and a mineral/vitamin tub.

Ann has learned to use nutrition and exercise to reduce birthing difficulties. "The most important part of pregnancy health is animal weight," Ann says. "The does need to be in top shape at breeding time and kept at a healthy weight during early pregnancy to minimize birthing problems. Exercise and browsing late into the season is important." The mature goats are housed in an open barn. Kids are raised with a CAE (Caprine Arthritis Encephalitis) prevention program. "I am proud of our small farm enterprise," Ann states. "Heifer International gave me my love of goats and now I am able to donate goats back to Heifer. The satisfaction comes in being in partnership with the goats, watching a young child try the cheese and wanting more, and hearing praises for the quality product from a market customer." ◆

THE LEARNING GUIDE

PEOPLE, ECONOMICS AND MARKETING

LEARNING OBJECTIVES

By the end of the session, participants will be able to:

- Identify ways in which each family member contributes to the farm enterprise and why each is important
- Identify all of the links in goat production and state why each is important
- List the benefits and costs (monetary, hidden and opportunity) associated with their goat enterprise

TERMS TO KNOW

- Gender
- Family
- Value Chain
- Budget
- Markets: local, niche, national
- Costs
- Profit
- Assets

MATERIALS

- Large poster of a blank activity profile
- Blank poster or newsprint for preparing a "Value Chain"
- Crayons, colored pencils or small sticky notes
- Copies of a Farm Budget template
- Pens and paper
- A sign that says "One Year Later"

ADVANCE ASSIGNMENT

Select four class members to discuss the skit and prepare to present it to the group. Give them a copy of the sample script to read.

TIME (May vary according to group)	ACTIVITIES
15 minutes	**Group Sharing** ■ The group will have met for 7 to 10 weeks at this point in the classes. Discuss with the group what has been the most helpful to them in this period of time.
20 minutes	**Get Everyone Thinking and Talking** Skit ■ See sample script at end of this Learning Guide of a theater skit about marketing and decision making. Actors *ad lib* some of the conversation once they know the skit. Add and change conversation to fit the situation within the community. ■ What did we learn from the skit? ■ How did the group go about making decisions?
20 minutes	**Doing an Activity Profile** ■ Use the blank Activity and Labor Profile poster. Decide on colors for women, men and children. ■ Divide into groups of all men and all women, maximum five per group. ■ Identify the tasks and show who is doing them. Discuss if this is true for all families. Discuss if anything needs to be changed. ■ Look at the activity poster and discuss how people feel about the work. What makes them happy to be contributing to the goat enterprise? What could be better? ■ Identify ways to acknowledge the work people do.
25 minutes	**Fill in the Access and Decision Table and Discuss (Same Group)** ■ What is the best way for families to make decisions? ■ How do you feel about the decisions as they stand on your own farm?
20 minutes	**Value Chain** ■ Review the Value Chain in the book and on a large poster make a Value Chain that represents the input and output of a potential goat project in the community. Begin by determining if it is a dairy goat or meat goat project.
45 minutes	**Budgeting** ■ Divide into pairs. Spouses should be pairs if they are both present. If a participant's spouse is not present, suggest that he/she review this with him/her at home. ■ Hand out budget forms and have everyone begin to work on their farm budgets. Each person will do an individual budget, but will discuss it with his/her partner. There should be several facilitators who can help get the activity started.

REVIEW-20 MINUTES

- What was useful?
- What was a surprise?
- What did not work out well?
- What do you know now that you did not know before?
- What can you do as soon as you get home?
- What practices will be difficult to do at home?

SAMPLE SCRIPT FOR GOAT MARKETING AND DECISION MAKING

Feel free to modify according to your context (change names, etc.)

Actors: Four people are gathered sitting on the ground or in chairs.

Surya: I called our meeting today because as officers of our goat group I think we have to look again at what we are trying to do. We all thought we wanted dairy goats. But what have they brought to us?

Sita: Well, I have milk for the family, but the concentrate I have to buy almost equals the value of the milk. And the other thing is that I am spending more and more time with my four goats and less time with my children and husband. If only I did not have to milk twice a day!

Mahendra: I tried to sell milk and cheese at the farmer's market, but people here are not accustomed to goat milk and they certainly have not developed a taste for goat cheese. So I usually take it all back home. Of course we use it, even the soured milk for cooking—but it is not what we expected. What people want is goat meat, not goat milk.

Dilip: Can we be courageous enough to change our direction and to start again with a meat-type goat? Even the cross-breeds we are using can be adapted to this.

Surya: It will be difficult, but I am ready to try it. If goat meat is the niche market, then we should go that way.

Sita: I agree.

One of the actors holds up a sign that says: "One year later." The four people are gathered and are counting money.

Mahendra: Can you believe it was only a year ago when we were complaining about no income from our goats? Now look at us we even have money to give micro-credit loans to other farmers and get them started with meat goats.

Dilip: The demand is greater than we thought. Our cooperative is growing and the local abattoir is so grateful for all the extra customers he has since we have been selling goats for meat.

The End

THE LESSON

PEOPLE, ECONOMICS AND MARKETING

INTRODUCTION

Goat keeping is a small business for both full-time and part-time producers. It is usually a family business, with all members of the family having important but different responsibilities. Record-keeping, and analysis of costs and benefits are essential for success. Goats are rarely the only enterprise on the farm, so producers must calculate how the goats (or more goats) will fit in with other activities. For example, if the goat herd increases from four to 20, will there be enough forage, pasture and other resources to care for them? Are there enough people to do the work? The individual farm is a single business within a wider network of similar farms and market opportunities.

ECONOMICS AND BUDGETING

Before deciding to raise goats, identify your goals. Are you raising goats for family consumption – for milk and meat? Are you raising them to market milk, or to market live goats? Is there a niche market for other value-added products?

Determine what resources are available. Do goats fit well into the farm plan? Can some of the forages be grown or will they be purchased? Is it possible to trade labor with a neighbor in exchange for milk or manure or offspring? Can enough saplings be cut to make a zero-grazing unit and a fence for an exercise yard for the goats? Can supplemental feeds be easily obtained? For dairy goats, what about the wood needed to build a milking stand, or money to buy a stainless steel bucket? How close is the nearest animal health technician, and will they visit a sick animal?

A budget allows one to view all of the costs of producing a product whether it is a liter of milk, a live meat goat or a kilo of cheese. Some costs are obvious, like the price of a kilo of concentrate. There are hidden costs like the value of time invested in goat raising, or borrowing a neighbor's wagon or truck to take animals to market. And there is the "opportunity cost," the value of something if it were not used for goats. For example, if planting a field of alfalfa, the same field cannot be used for maize. The value of the maize is the opportunity cost. The market value of the alfalfa may be less than the maize, but after feeding the alfalfa to the goats, and selling their milk, more money may be made than simply selling the maize. The budget helps identify which farm activities provide the most income.

Profit is the difference between the cost of producing something, and the benefit from it. If it costs $10 to raise a meat goat to 40 kg, but the goat can be sold for $50, the profit is $40 per goat. The money can be reinvested in the farm by buying more goats, or using some of it to improve the house, pay school fees, buy clothing, food, or materials to make farm sausage.

ACTIVITY AND LABOR PROFILE

On small family farms, available family labor can limit goat production just like available land and water. Each farm should perform an activity profile to understand who currently performs which farm chores, and who will do so in the future, as the herd increases. Is there enough family labor? Will the children be able to help without missing school? Will there still be time for other chores like caring for the children, cooking, fetching water and firewood and working at off-the-farm employment? Is there cash to hire additional labor?

Here is a sample chart to understand "who does what" for the family-based goat enterprise. It is also useful to follow up with a "who does what" for the entire farm household, including off-farm labor, domestic chores that are unpaid (cooking, cleaning, home repair) and community meetings where information is shared.

ACTIVITY AND LABOR PROFILE									
Person	Child Care	Cooking Cleaning	Feeding Animals	Milking	Processing Milk Products	Marketing	Training Meetings	Care of Household Garden	Total Hours Per Day
Adult-male									
Adult-female Senior									
Adult-female Junior									
Child-male									
Child-female									
Elderly-male									
Elderly-female									
Hired-male									
Hired-female									
Neighbors-male									
Neighbors-female									

There are two separate categories for "adult women": senior women (mothers-in-law) and junior women (new brides or daughters-in-law). In some areas of the world, the greatest challenge is getting the young wives to training, since this is seen as the privilege or reward for the older women who dominate the younger.

The charts can also help determine which member of the family or staff attends training sessions. We all know that the person who does the work will benefit the most from training. However, in some places it is not traditional for women to travel, or receive education. It is important that both women and men who work with livestock have time to attend training and group meetings. Sometimes special arrangements need to be made for child-care and tending to the animals if women are away for training, especially if it is for several days. The activity analysis can highlight issues on the farm, such as too many tasks for one person. Solutions such as sharing chores, or investing in time-saving equipment like forage choppers or wheelbarrows, can be explored.

ACCESS AND DECISION TABLE

After the Activity and Labor Profile, the Access and Decision Table on the next page shows what types of benefits come from the goats, and who in the family is involved in the activity and who receives the benefits.

For example, in a Tanzanian village, only the men can decide (D) when to slaughter a goat for meat although the women have access (A) to the meat after it is cooked and served to the men. Or in a Fulani group, goat milk belongs only to the women, and they decide whether to serve it to the family or sell it in the market. The men can consume the milk that is given them by the women, but also decide when to wean and sell the kids.

The person who feeds, milks and cares for the goats needs both proper training and good incentives. Often, if women can retain use of the cash from milk or meat sales, they are more efficient and interested in their work. On the other hand, when women are viewed as free labor in the family, but the husband takes all the profit for himself, the quality of work declines and production is lower.

When we compare the different perceptions of men and women, and discuss how families can improve their cooperation and mutual support, we have a better understanding of the roles of each person in the goat enterprise. What incentives do those working with the animals have for excellent work? In general, the more incentives for workers (either family members or paid labor), the greater success. Incentives can be cash, training opportunities, offspring, use of profits and even public recognition.

ACCESS AND DECISION TABLE							
Person	Meat	Milk	Manure	Cash	Hides	Gifts	Other Benefits *Training, Public recognition, Personal satisfaction*
Put A for Access and D for Decision							
Adult-male							
Adult-female Senior							
Adult-female Junior							
Child-male							
Child-female							
Elderly-male							
Elderly-female							
Hired-male							
Hired-female							

BUDGETING

First list the overall goals for the goat enterprise. Then use the chart on page 155 to list your resources, where you will get them and at what cost. The second column is for benefits, such as the price you can sell each liter of milk. The value for the household is the cost that milk consumed would have otherwise cost, plus other "intangible" benefits such as healthier children from better nutrition. Include all benefits from the goats, including manure and sale of kids for meat.

Other items can be added to make an expense and income budget for the goat enterprise. Also, do a yearly budget for the whole farm, including all crops, trees and off-farm employment. Work with an extension person to make this fits your personal situation.

THE VALUE CHAIN

The Value Chain is a diagram of all the "links" in a certain economic activity, like goat milk production. All of the links, from acquiring the goats, feeding, health care, processing, transportation and markets, must function well for each participant to be successful. In addition, drawing or listing the links in a Value Chain allows you to see where the greatest profit lies, and suggests ways for producers to increase their portion of the income generated throughout the chain. The producer who sells a raw product like fluid milk or live goats may receive only a tiny fraction of the value from the goats. Greater cooperation with processors or forming a producer's co-op can increase profits that benefit everyone. For example, if a group of farmers turn their milk into cheese, and

YEARLY FARM BUDGET FOR GOAT ENTERPRISE				
Item	**Where Available**	**At What Cost (Expense)**	**For What Benefit**	**Value for Household (Income)**
Four goats	Two from Heifer, two to purchase	Two offspring for passing on the gift–$80 to purchase goat kids		$
Zero grazing unit and fencing	Can cut saplings from roadside or get lumber from market	$		$
Forages	Have some forages and will plant others			$
Concentrates and mineral salt	Purchase from local market	$		$
Milking equipment	Build milking stand – get buckets and other equipment from market	$		$
Health care		$		$
Sale of milk and milk products		$		$
Sale of meat and meat products		$		$
Sale of kids		$		$
Exchange of milk and manure for labor		$		$
Manure for Garden		$		$
Milk and meat for family		$		$
Hired labor		$		$

then contract to sell a defined quantity and quality of cheese to a distributor, they can command a higher price per kilo than each selling on their own at the local market.

The Value Chain shows that the market needs to be present before increased production occurs, not afterward. Nothing is more discouraging to goat producers than having a good product, like milk, and not being able to sell it or selling it for less than the cost of production.

The advantage of the Value Chain approach is seeing how goat production fits into the "big picture" of farming, income production, family benefits and rural development. Each link is important and must be addressed before goat production begins. One weak link can undermine careful work in all of the other areas.

The links in this particular Value Chain are: Inputs, Supporting Services, Production, Marketing and Sales and Outputs (Results, Environment).

Inputs

- What are the raw materials for production? (animals, land, water, labor, goat feed, crops, equipment and buildings) Are they easily available, or do new sources need to be arranged?
- Do you have enough cash to make initial purchases?
- What is the current work activity within the family of farming and off-farm work?
- Does the family have time, knowledge and the ability to perform the labor needed?
- How will both women and men participate in the labor?
- Who will participate in decision-making?
- Will the benefits from the enterprise be shared?
- What are the risks to farmers for getting into (or expanding) goat production? What can alleviate some of these risks?

Supporting Services

- Is training available for all people who will be working in the goat enterprise? The training may be done by extensionists, veterinarians, Community Animal Health Workers (CAHW) or project leaders. Women may need separate or additional facilities for training or they may attend the same training groups as the men, depending on the cultural norms. Arrangements for child care will mean more women can participate.
- Are micro-credit and other loans available to help with additional costs of getting the goat enterprise started? What government subsidies exist in agriculture or food processing that might affect goat production? (subsidized feed, fertilizer, cheese producing facilities, cooperatives) Are these subsidies likely to continue into the future?
- Are livestock medicines and veterinary services available?
- Culturally supported sharing of benefits has allowed societies to survive and thrive but some individuals can take advantage of it, and threaten the profitability of an enterprise. Solutions may include bank accounts, offering jobs to needy family members, or helping them with formal loans. Once income from the goats is established, farmers need to decide how to handle requests from their extended families for free milk or cash. This has been the downfall of many goat projects, and requires advance planning.

Processing, Retail Production and Marketing

Products from the goat farm may include live goats, meat, milk, cheese, yogurt, manure and skins. In a household system processing may be done at home or with a few neighbors joining together. However, "scaling up" commercially or expanding can increase income.

Raw products like fresh milk and live goats can be sold to local consumers, but often are bought by middlemen who process them in distant places. Processing milk into cheese or a live goat into retail cuts of meat is called "adding value" because consumers pay more for it. Adding value through processing is where the greatest profits lie, but farmers often are not able to do this without good planning.

Explore markets for both raw and processed goat products. Are there special markets in the village or town for milk, for meat or for kids? Visit different markets, and learn about the range of prices. It is difficult to negotiate a good price when you do not know its value to the final consumer. A market may be closer than imagined. Neighbors may want to purchase a product or to trade labor for products. If better markets are farther away, organize a group of farmers to transport a large batch, and share costs.

Farmer groups can also organize networks to share price information from different regions of the country, through cell phones, radio or internet announcements. In many countries, government or farmer co-ops are responsible for "price transparency" as a way to help farmers negotiate a fair price.

If there is no market, perhaps one can be created. Plan this carefully. Consumers must be aware of the product and its value. In some places, urban consumers now prefer goat meat because of health preferences for leaner, lower calorie products that are low in cholesterol. For better marketing, producers keep quality and consistency of products high. Feedback from consumers is an important link in this process.

There are many resources to help people market products. Start by talking with local cooperatives and neighbors. Look around for current markets that people are using, especially where women are participating. Be aware of the difference in price for high quality products over those of lower quality. Attractive packaging also helps attract customers.

Local markets often become centers where people not only sell their products but exchange information, learn new skills, socialize with one another and develop a strong community spirit.

Environment

The environment shows the big picture in developing a goat enterprise. We are using this in its broadest context and thinking of it in a local and global context. Some of the things we will consider are environmental change, cultural norms, market demands, world trade agreements and political leadership.

What rules and official inspections regulate animal production, health treatment, use of chemicals, processing, marketing and slaughter? How are necessary licenses and inspections obtained? What is the cost? Learn if products like goat milk need to be tested or inspected before sale.

YOUR OWN VALUE CHAIN

The links in this Value Chain are an example, but they may not fit your situation. You can keep those that are useful and develop other links as appropriate.

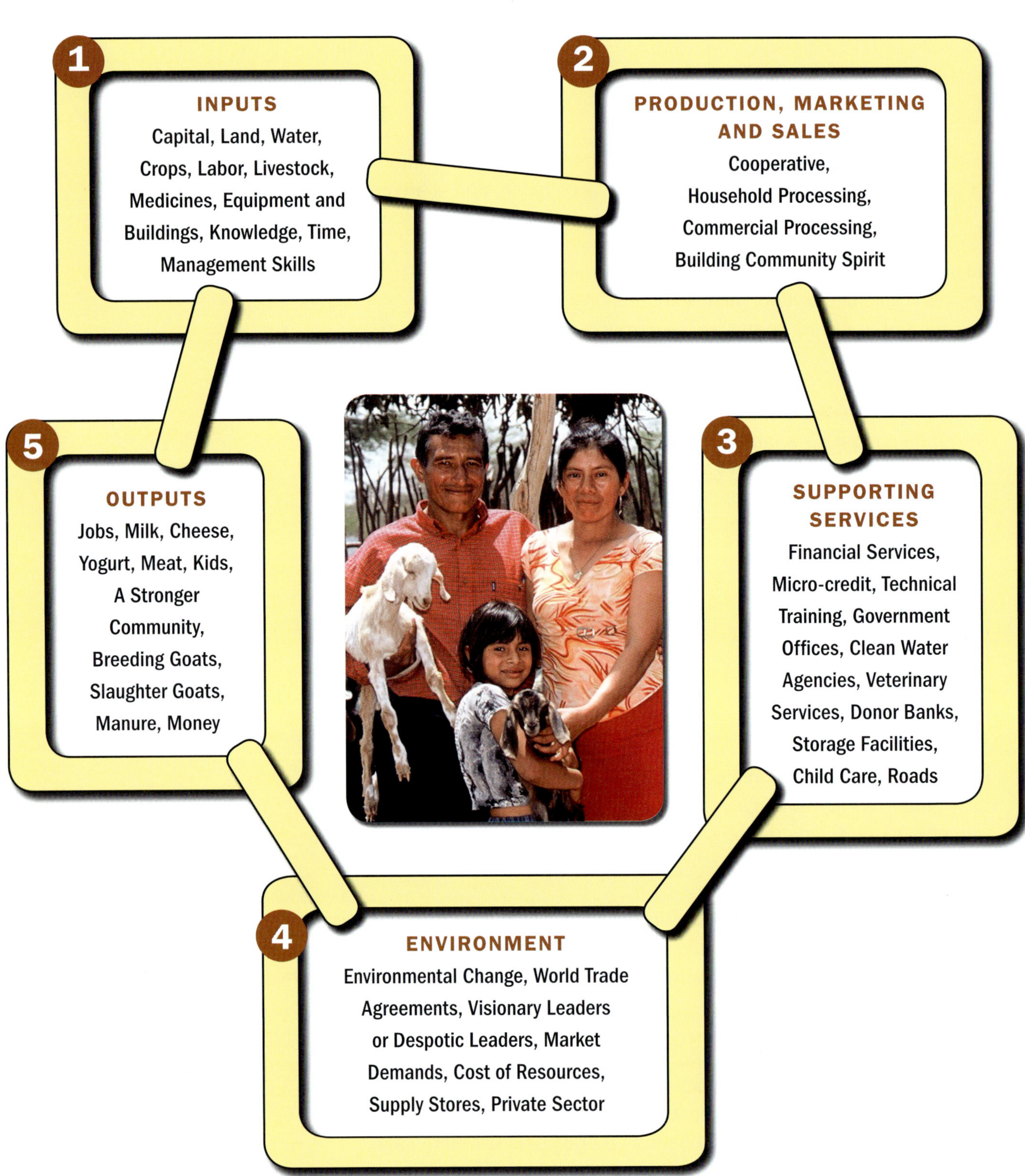

Women and Goats in Morocco

▲ Moroccan woman with goats.

What does gender have to do with raising goats? Isn't the goat the same, whether the owner is a man or woman? Oddly enough, it does matter whether the owner or manager is a man or woman, especially in developing countries, where men and women often have very different access to resources like land, information and even time.

Morocco is an example of a country where goat keeping is an important part of the rural economy, and where gender matters in credit, extension programs, decision-making and marketing. As the transition from subsistence to market economy proceeds, women often lose the benefits from goats unless policy and development practices pay attention to gender.

In the subsistence economy in Morocco, there is a gender division of labor, but some roles are flexible. Typically, men take the goats (and often sheep) out to the parcours (pasture) to graze during the day, but women or children are often found tending goats as well. During milking season, the women milk the goats, and use the milk to feed the family. In times of surplus, they make cheese for home consumption or sale. The men sell the cheese to neighbors or through informal markets, since women are supposed to stay in the home. Although it is a Muslim country, and the income that a woman generates is supposed to be hers alone, in practice the money does not always return to her. When goats are sold for meat or slaughtered for home consumption, it is the man's decision. Note that goats for meat are important for religious holidays, weddings, funerals and other social occasions.

As a Muslim country, traditional gender roles are important, especially in rural areas. The woman has responsibility for managing the home, and the husband for everything else, including land, agriculture, transportation, purchases and joining organizations. Although women provide labor in agriculture, for the goats and the crops, most men do not consult their wives on decisions affecting the land, crops or animals.

Morocco like many countries is trying to commercialize goat milk production to provide steady income for rural people. However, once goat milk becomes a valuable commodity, men begin to take over what has been women's domain. Women, who used the money they once generated from milk and cheese sales, now must ask their husbands for cash to buy household items, causing domestic conflict. Children's nutrition may suffer if all the milk is sold, without keeping any behind for home consumption.

One solution in Morocco has been the creation of women's cooperatives for dairy goats, in which the income from the goats goes into the cooperative's

Continue on page 160

Continued from page 159

bank account, which only the women can access. The co-op can access credit which the women cannot do individually, since they lack ownership of land to use as collateral.

Another approach is to include gender sensitivity and analysis in co-op development, and to include all members of the family when planning the project. An explicit discussion of gender roles in the family means that husbands and other men can understand the position and responsibilities of the women, and choose to be more supportive, so the entire family benefits.

Another program ensures that women as well as men will benefit from livestock extension. A survey revealed that women's priorities were health care for their children and reproductive health for themselves. A training program in goat health and husbandry was developed, based on women's existing knowledge and interest in human health. For example, treatment of diarrhea in human babies and goat kids is similar, and understanding one can help women understand the other.

Another benefit of goat projects for women has been technical training for their husbands. In Morocco, although rural women are more likely to be illiterate than men, they are more open to change and trying new technology. Modern goat raising using vaccinations, dewomers and supplemental feed are accepted by women more easily than their men folk. However, the men eventually accept the new practices when they see the improved health of the animals.

Goat husbandry will continue to provide livelihood for rural people in Morocco. During the transition from subsistence to commercial production, it is essential that development workers remember these simple rules about gender:

1. Gender means both men and women are included, not just special projects for women alone.

▲ *Berber rug made from goat hair.*

2. Goat projects are popular with both men and women, and are a good opportunity to discuss how men and women work together to benefit the whole family.
3. Goat husbandry can be a successful means of organizing women and increasing their income, but only if they can keep the income they generate, which requires their husbands' support.
4. Development workers must understand the different opportunities and constraints for men and women farmers, especially the heavy workload of rural women.
5. Extension for illiterate women must be in the local language. Avoid technical jargon, rely on pictures rather than written materials, and build on their previous knowledge and interests.
6. Extensionists and other animal health workers benefit from training in "Participatory Methods," which increases the impact for training rural men and women.

Gender Equity means stronger and healthier families today and tomorrow. ◆

THE LEARNING GUIDE

HUMANE TRANSPORTATION AND SLAUGHTER OF GOATS

LEARNING OBJECTIVES

By the end of the session, participants will be able to:

- Know how to properly care for an animal prior to slaughter
- Know how to slaughter a goat for family use or for the market
- Understand local regulations regarding goat slaughter and marketing
- Know how to preserve the skin of a goat

TERMS TO KNOW

- Offal
- Slaughtering Cradle
- Carcass
- Viscera
- Sanitation
- Fasted
- Fell
- Pluck

MATERIALS

- Goats for slaughtering (Be sure they have been fasted at least 12 hours.)
- Wood for building slaughtering cradle and hanging stick
- Hammers, nails (Wood can be pre-cut to shorten the exercise.)
- Sharp knives
- A place to hang the goat after slaughtering
- A saw for cutting through the sternum
- Buckets of soapy water and clean water for washing
- Shovel for burying offal (the viscera and trimmings of a butchered animal removed during slaughter)
- Salt for preserving goat skin
- A small barrel for making a drum

ADVANCE ASSIGNMENT

A clean area for slaughtering should be selected. Goats for slaughtering should be selected in advance so they can be fasted at least 12 hours before slaughter. Decide how the meat will be used. This lesson may be divided into several sections.

- Discussion and preparation of necessary materials
- Slaughter of the goat
- Preparing the skin (hide)
- Making a drum from the skin

TIME (May vary according to group)	ACTIVITIES
20 minutes	**Group Sharing** ■ Discuss how people are feeling about the slaughtering activity. ■ Talk about the sacredness of life and animals as an important source of food.
15 minutes	**Presentation on Local Regulations**
30 minutes	**Make a Slaughtering Cradle and a Hanging Stick**
30 minutes	**Observe the Preparation and Slaughter of a Goat**
1 hour	**Practice Slaughtering a Goat**
1 hour	**Practice Preserving the Skin of the Goat**
20 minutes	**Bury the Offal**
1 hour	**Distribute Meat or Prepare a Festive Meal for Members and their Families**
45 minutes	**Make a Drum (Optional)**

REVIEW-20 MINUTES

- What was useful?
- What was a surprise?
- What did not work out well?
- What do you know now that you did not know before?
- What can you do as soon as you get home?
- What practices will be difficult to do at home?

THE LESSON

HUMANE TRANSPORTATION AND SLAUGHTER OF GOATS

INTRODUCTION

It is important to know how to slaughter animals for home consumption. Slaughter of any animal should be done in a humane manner by a properly trained individual.

A goat carcass contains bone, muscle, and fat. The muscle is primarily protein (composed of amino acids) and water. Goat meat protein is said to be "high quality," i.e., it contains substantial quantities of the essential amino acids, in appropriate ratios, for human dietary needs. The composition of goat fat varies considerably from that of beef, pork or lamb. These differences account, in part, for variations in flavor between chevon (goat meat), beef and lamb. As a source of human dietary energy, goat fat is equal to beef or pork fat; and goat fat is approximately about equal to beef fat in degree of saturation.

SELECTING GOATS FOR SLAUGHTER

In a meat goat enterprise, the goats are raised for meat and will be slaughtered at the appropriate age to meet family needs and market demands.

In a dairy goat enterprise, goat kids, poor producers, surplus breeding stock and older animals may be slaughtered.

When there is insufficient feed for animals because of drought or other environmental circumstances they will be slaughtered for human diets. On occasion animals with genetic defects will also be culled. If abnormal conditions are evident, further inspection by a veterinarian or technician is advisable.

WARNING

Goats with infectious diseases like brucellosis, tuberculosis and anthrax should never be slaughtered for human consumption.

TRANSPORTING ANIMALS

- Transportation vehicles should be checked for safety and adequacy
- Vehicles should be clean
- Goats are normally fasted for 12 hours prior to slaughter, but should continue to receive clean water
- It is recommended that, when possible, the animals be transported to the place of slaughter and then fasted

- Adequate space should be provided during transport to allow animals to stand freely and in a normal posture
- Segregate animals by different size and age
- Time in transit should be considered. If the journey is more than several hours, there should be a provision for off-loading, rest, exercise, feed and water
- Animals should be transported in the cool of the day
- If animals are walked to market for sale or to an area for slaughter, rest, feed and water should be considered
- Always treat animals humanely

HANDLING OF GOATS PRIOR TO SLAUGHTER

Slaughter should be performed by a properly trained person. Keep equipment in good working order. Complete slaughter as quickly as practical to minimize animal anxiety. Only healthy animals should be slaughtered for food. No animal that has died from unknown causes should be used for human consumption. Be sure withdrawal time of all medications has passed. Provide plenty of clean drinking water. Place animals in a clean, well-bedded area. Do not excite animals prior to slaughter. Do not pinch hide or beat with a stick because you will stress the animal, bruise the meat and it will spoil more quickly. Slaughter in the cool of the day. If the meat of the animal is not to be eaten in 24 hours and you do not have refrigeration you will want to preserve the meat by salting or drying.

Equipment

- Slaughter in a clean area, preferably with a concrete floor, using a table or slaughtering cradle
- Make a hanging stick or put a rope through the Achilles tendon and leg bone above the hock
- Make a Slaughter Cradle. Have plenty of cold water available.
- Keep all dogs away from area
- Having the goat on the slaughtering cradle helps keep the animal's carcass clean

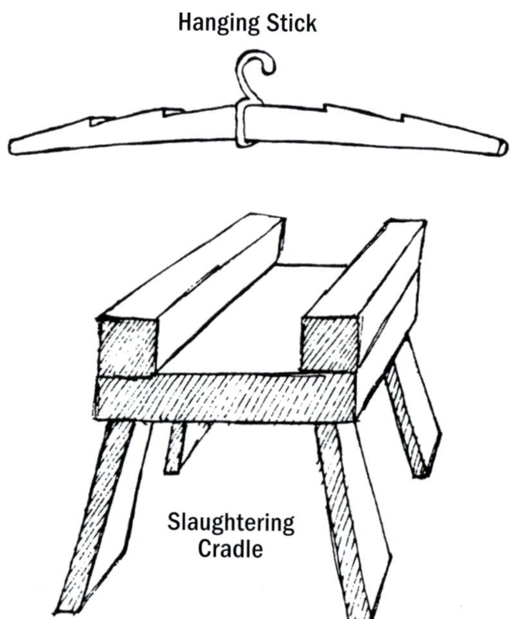

The slaughtering cradle should be large enough to fit a mature goat and additional pieces may be set inside for a smaller animal. The height should be comfortable for the user.

- 2 buckets filled with clean water
- 1 empty bucket for blood
- Hammer or Stunning Pistol (a stunning pistol is made for this purpose; it is not a firearm)

SLAUGHTERING PROCESS

1) Halter animal. Hold at about 6" distance.

2) Some people prefer to stun the animal first. Draw an imaginary cross between ears and eyes. Where the cross intersects is the place to strike the stunning blow. Use small blunt instrument like a hammer or a stunning pistol. The animal will probably not die, only be stunned.

3) Put the stick knife in below the ear and at the base of the jaw. Stick through to other side. Cutting edge should be pointed out. Speed is important. Let the blood flow freely. When you are sure animal is through bleeding, pull chin back until you can see neck bones. Sever head.

4) Let animal bleed a couple of minutes until it stops thrashing.

5) Put animal into cradle on its back, head down. Wait until reflexes have stopped.

6) Ring each hoof between hoof and dew claw with knife. Run point down back of each front leg. On rear legs run knife all the way up leg to rectum. Go just through the skin. As you begin to trim skin away, the dew claw becomes the holding point.

7) Separate skin from pastern area. Jerk hide down to hock holding onto dew claws.

8) Now jerk skin up to crotch. Do both sides. Be careful not to break the connective tissue called the "fell."

9) Separate tissue from severed line in crotch. Leave scrotal or mammary tissue in hide. Separate flap of skin from rectum to leg.

10) Front legs: do the same, jerking skin carefully to the body.

11) Draw knife across sternum from armpit to armpit. Go right under skin. Do not cut meat. At the point the knife crosses brisket, cut to throat and open. Free the entire wind pipe (trachea) up to the sternum.

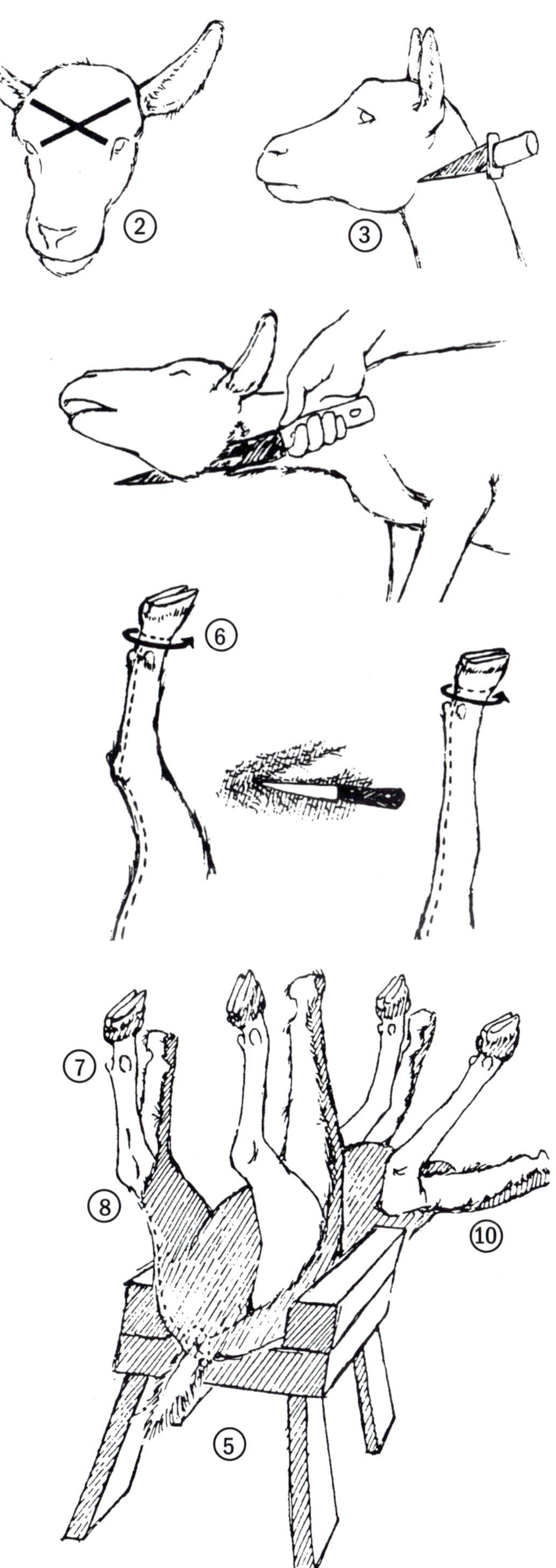

12) Begin working the skin back to the armpits.

13) The front end should be skinned, exposing throat, front portion of each shoulder and legs.

14) Once the hide has been separated from the legs at the break joint, sever the foot from the leg. (The break joint is only present in animals under a year of age.) Otherwise, remove foot at ball joint.

15) Take knife and open skin from sternum to crotch. When slaughtering a male animal let the penis hang and remove when you are ready to cut up carcass. This eliminates a blood deposit in the crotch.

16) Work hide back to each side using fingers and fist. Try to keep thin layer of connective tissue on the carcass. It is important to keep carcass intact. The entire stomach is now exposed.

17) Put the hanging stick through the leg between the Achilles tendon and leg bone above the hock. Or put a rope loop through each rear tendon so the legs are tied together.

18) Hoist animal off of cradle so it is hanging. Be sure carcass is secure.

19) Cut with knife from throat to sternum. Use a saw to cut through sternum. Chest cavity is exposed.

20) With fingertips and fist, work hide over rump. When approaching the tail, grasp tail base and pull skin off.

21) At this point strip back with fingers and knuckles trying to keep all connective tissue on carcass, going right down backbone. Again, keep subcutaneous muscle attached to carcass and not hide. Pull off. Lay hide on rack. You will work with hide later.

22) Go to front of animal. Wash carcass with clean water.

23) Put in knife at crotch, cutting side toward you. Put finger behind dull side of knife and push finger and knife down toward navel and eventually to sternum. Guts will fall forward.

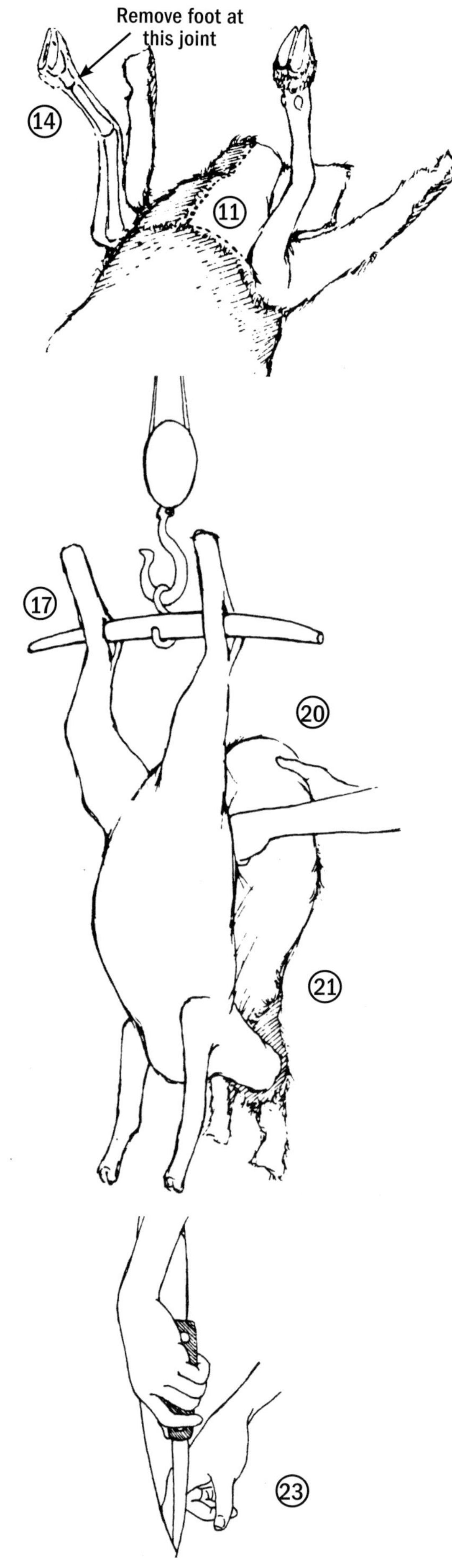

24) Ring the rectum with knife. Tie it off. This is called the "bung." Reach into the crotch and draw the rectum through pelvic arch.

25) Pull bladder free and let this go with guts. Leave kidneys and kidney fat in the body cavity.

26) Approaching abdominal cavity, spleen will be on left hand side. Pull loose—discard. At this point the liver will be visible on the right. Grab base of gall bladder and strip from liver, leaving liver in body cavity.

27) The bottom of the esophagus should be visible. Tie this off. Cut this and guts will fall loose on the ground.

28) Beneath the kidneys and liver, locate the diaphragm, a distinct red and white line. This band separates the thoracic cavity from the abdominal cavity. Run knife along muscular strip (red and white line) and cut to sternum on each side. Grasp in center and pull away from muscle supporting liver. Do not take out.

29) Heart, lungs, windpipe and esophagus (pluck) will be in the material you are dragging away. Remove pluck, opening the throat the full length of the chest cavity.

30) Thoroughly wash carcass inside and out with clean water.

31) Remove tongue from head and any other desired parts of head. It is not recommended to use brains or spinal material because of concern about Bovine Spongiform Encephalopathy (BSE), commonly known as Mad Cow Disease.

32) Keep the meat in a cool place or throw cold water on it periodically. If you are going to leave hanging for any length of time, wrap in a clean cloth.

33) Cut in desired pieces. Edible by-products would include liver, heart, tongue and kidneys. Any organs or tissue that appear abnormal should be discarded. If there is extensive abnormal tissue, the entire carcass should not be eaten. The small and large intestines can be salvaged and used for casings for sausage.

34) All inedible viscera should be disposed of in a sanitary manner and buried deep beneath the surface of the ground to prevent dogs from spreading tape worms.

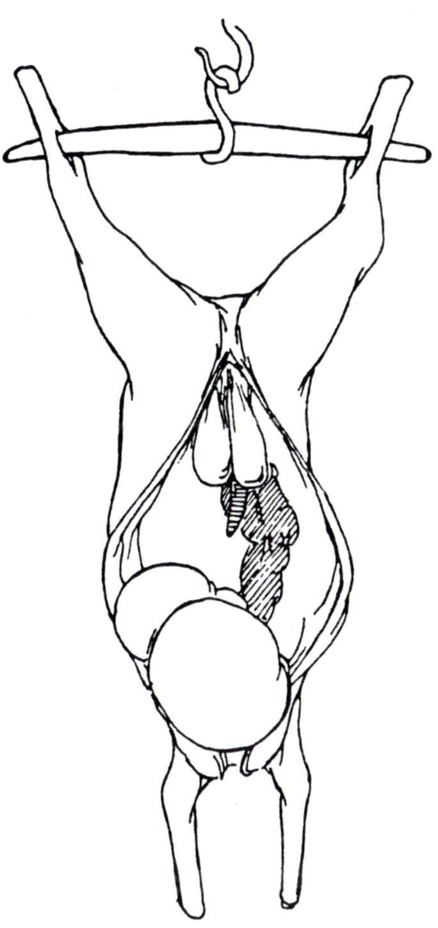

WARNING

Take special precaution when handling spinal cord or brains.

PRESERVING THE SKIN (HIDE) OF A GOAT

After slaughtering the goat use a flat tool to scrape the excess tissue from the non-hair side of the skin and make sure there are no pieces of meat left on the skin. After that, put salt on the entire area that you have just scraped. Rub the salt into the skin. Tie the skin taut and dry in the sun or in a room which has the same temperature as the sun for about seven days or until it dries completely. The skin will become hard as it dries. After it has dried work the skin with your hands until it is pliable then it can be washed and worked again.

Skins can be used for clothing, for bags and for drums.

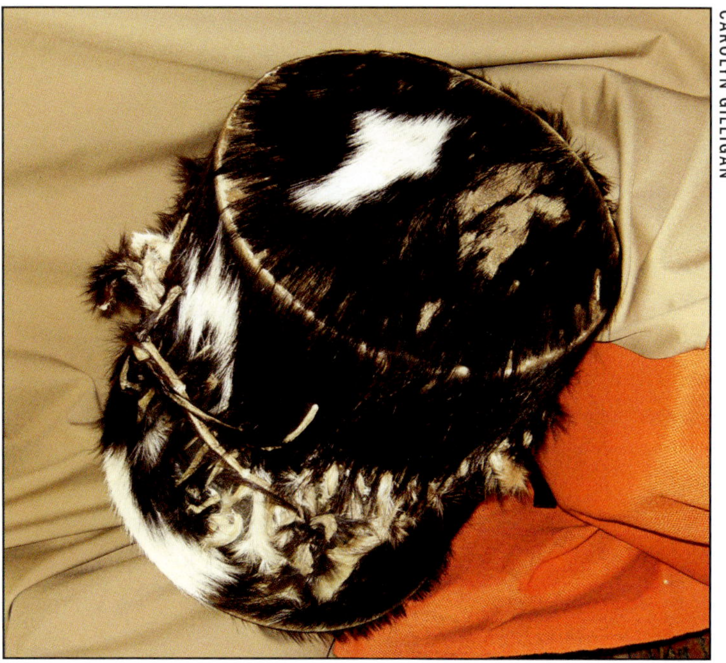

CAROLYN GILLIGAN

▲ *Goat skin drum*

Newfound Dignity and Youth

▲ Joyce Fleki

Joyce Fleki, 72, a member of the Masakhanyise Project in the Eastern Cape Province of South Africa, said that not long ago, her life held little promise. "Before I joined this project I was helpless. No one wanted me. I was very thin and poor. I used to walk up and down the village begging for food."

But her life has changed dramatically. "I am very active now," she said. "I now have dignity. I am respected by my children and the community in which I live."

Before starting her work with Heifer South Africa Mrs. Fleki had no assets and very little hope, she lived with her children and saw herself as a burden to them.

In 2004, she received two pregnant Boer goats, a South African breed of meat goat. Now she has 18 Boer goats, and she has honored her commitment by passing on the gift of two goats to another family, who she also helped to train.

"The first goat I sold was a male goat, which I sold for 800 rand (US$120) and bought myself a kitchen cupboard," she said proudly. "For years my dishes used to lie on the floor because I could never buy anything to store them in. I feel like a powerful woman because I have a cupboard for my cups and plates."

Mrs. Fleki's village consists of approximately 300 dwellings, varying from mud and stick-built round houses called rondavals to more secure cement block two- to four-room houses. The area is typically rural, with very limited infrastructure.

Families get water from communal taps situated strategically within the village. Electricity became available only a few years ago and because of its high cost is only used for lighting. Villagers still cook food on open fires. Up to 90 percent of the village's residents are unemployed, and there is little hope for young people.

But the achievements of Mrs. Fleki and the community group are an inspiration to fellow villagers, who have seen very few successful projects in their lives. Mrs. Fleki said the project has not only made her feel successful, it has also made her feel young again. "Look at my skin," she said, holding up her arms. "I am young and healthy now!" ◆

REVIEW AND CELEBRATE/CERTIFICATE

Share in the preparation of a goat barbecue to which families will be invited. Select individuals to present skits about what they have learned. Give certificates to those who have completed the course.

GLOSSARY

Agroforestry
A system of land management involving the growing of trees with food crops and animal husbandry.

Anestrus
When a female does not come into heat (estrus).

Anthelmintic
A drug used for killing parasites.

Artificial Insemination
The process of breeding a female goat with semen collected in a container from a male goat and then frozen or transported fresh to the doe(s). This can increase the number of does an improved buck can breed.

Biodiversity
A naturally wide variety of species of plants and animals that interact with the environment.

Billy
The slang word for a male goat.

Body Condition Scoring
A monitoring tool for assessment of the goat by palpation to help measure the adequacy of the nutritional condition of the goat.

Browse
Plants with woody stems such as shrubs and small trees.

Buck
An uncastrated adult male goat.

Buckling
A young male goat.

Budget
A written document for financial planning.

Bung
A term used for tying off the intestines or other parts of the carcass during slaughter.

Bush Goat
A goat produced from a sire and dam of unknown breeds. Also called local goat.

Caprine
Of or relating to goats.

Carcass
The body of a slaughtered animal after the skin and internal organs have been removed.

Castration
The removal of the testicles or crushing of the spermatic cord to make a male animal sexually sterile.

Cellulose
A major component of plant cell walls and a complex carbohydrate that human enzymes cannot digest. Microbes in the rumen of ruminant animals ferment it to yield energy.

Crude Fiber (CF)
A nutritional measure of cellulose and ligneous material.

Chevon
The meat of a goat.

Colostrum
A thick, yellowish liquid produced by the doe just prior to kidding and during the first few days of lactation. Colostrum contains immunoglobulins (antibodies) to help protect the kid from infectious diseases.

Compost
A mixture of decayed organic matter that may include animal manure. When left to decompose, it becomes nutrient-rich matter that is used for fertilizing and conditioning land.

Concentrate
A highly digestible prepared feed that is high in protein and/or energy and low in fiber.

Cost
The price paid to acquire, produce, accomplish, or maintain anything.

Crude Protein (CP)
The percentage of protein in feed and forage.

Crossbred Goat
A goat produced by mating individuals of different breeds.

Cud
A portion of food that a ruminant returns from the first stomach to the mouth to chew a second time. It is primarily roughage (fibrous material) and fluid.

Dam
A female parent.

Dehorn
Surgical removal of horns of older goats.

Dehydration
Lacking adequate water and salts in the body for normal functioning.

Dewormer
A drug administered for killing internal parasites.

Disbudding
Cauterizing the horn buds on young goats (3 days to 2 weeks) to prevent horn growth.

Doe
An adult female goat or other mammal.

Doeling
A young female goat.

Drug Resistance
Microbes or parasites with the genetic ability to live even after exposed to dewormer or antibiotic.

Egg/Ovum
The reproductive cell produced by the female, which after fertilization by a male's sperm is capable of developing into a new individual.

Energy
The nutrient, obtained either from fats or carbohydrates, that serves as fuel for the animal's body.

Ejaculation
The discharge of semen from the male's penis.

Escutcheon
The area under the tail on does forming an upside-down "U" which includes the anus, genitalia and rear udder attachment.

Estrous Cycle (adjective)
The time from one heat or estrus to the next. This is usually 18 to 22 days in goats.

Estrus (noun)
The time a doe is in heat: 24 to 72 hours.

Extensive Production
A husbandry system of free grazing and browsing under the supervision of a goatherd or within an area defined by fences. Shelter may be provided at night and during rains.

FAMACHA Chart
A testing system which utilizes an eye mucosa color chart to determine the level of anemia in goats. The color determines the anemia or blood loss caused by the blood-sucking roundworm *Haemonchus contortus* (also referred to as wireworm or barber pole worm).

Family
Traditionally defined as a group made up of parents and children and biologically related individuals. It may be more liberally defined as a group of individuals, usually related by blood or marriage, living in one place and contributing to the well-being of the household.

Fast
To deny a goat feed for either medical reasons or in preparation for slaughter.

Fell
A thin tough membrane, covering the carcass directly under the hide.

Fertilizer
A substance (manure or a chemical mixture) used to make soil more fertile.

Fetus
A young animal developing inside its mother's uterus before birth.

Fiber
Coarse plant material made up of complex carbohydrates (primarily cellulose, hemicellulose and lignin) that cannot be broken down by mammalian enzymes but can be broken down (fermented) by microbes. The microbes in the rumen of a goat break down cellulose and hemicellulose to yield energy.

Flushing
Increased feeding, especially of energy and protein feeds, several weeks before breeding to increase the number of eggs released and the chance of twins/triplets.

Forage
Plant matter eaten by animals that provides fiber, energy and other nutrients.

Forbs
Broadleaf, herbaceous (non-woody) plant species.

Fodder
Cut feed, pasture or browse that an animal eats, sometimes called "green chop."

Gender
The socially defined roles and responsibilities of men and women, which are not biological and change over time, and between cultures.

Gender Equity
The principle and practice of fair allocation of resources, benefits, work, leadership and decision-making to both women and men.

Genotype
The genetic makeup, as distinguished from physical appearance, of the animal.

Gestation Period (Pregnancy)
The length of time between conception and birth (kidding). This is usually 150 days for goats.

Goatherd
A person who cares for and herds goats.

Grasses
Herbaceous plants that have cylindrical stems and flat blades rather than broad leaves.

Green Chop
The term used in East Africa and other places to define plant material fed to animals, usually by cut-and-carry.

Hay
Harvested and dried grass or forage fed to livestock.

Herbaceous
Relating to or having the characteristics of an herb, or non-woody plants.

Household Goats
Goats kept for the primary livelihood of the family.

Humane
Characterized by compassion and sympathy for people and animals, actions to prevent suffering or distress.

Husbandry
The science, skill, or art of raising food crops or animals.

Hybrid Vigor
A positive result in cross-breeding whereby the offspring is more productive than either parent.

Immunoglobulins/Antibodies
Proteins that help an animal resist and fight disease. Kids receive these proteins through the doe's colostrum.

Infection
An illness caused by microbes.

Infertility
The persistent inability to achieve conception and produce offspring.

Insecticide
A drug that is applied to destroy external parasites like mites, flies and ticks.

Intensive Production
A husbandry system that provides permanent housing and in which all food is brought to the animals. Zero grazing is one of the intensive systems.

Intramuscular Injection
An injection of a drug or other substance into a leg or neck muscle.

Intravenous Injection
An injection of a drug or other substance into a vein.

Kid
A goat less than six months of age.

Kidding
The act of giving birth.

Lactation
A period of time when female goats or other female mammals produce milk.

Legumes
Plants and trees in the pea family. They have bacteria on their roots that allows them to convert nitrogen in the air into high protein forage or feed for livestock.

Life Cycle
The process of birth to death for a living organism.

Liver Fluke
A type of parasite that invades the liver.

Lochia
A bloody discharge after kidding that lasts up to 14 days.

Markets
Local, national or international places organized to buy and sell products. For goats, these products would be milk, meat, cheese or live goats.

Mastitis
Inflammation of the udder, usually caused by bacteria, such as streptococcus or staphylococcus organisms.

Microbe
An organism or particle so small (microorganism) that it cannot be seen with the naked eye. Many diseases are caused by microbes, which include bacteria, viruses and protists. Other microbes, like those in the rumen, help the animal.

Nanny
A slang word for mature female goat.

Navel
A depression in the middle of the abdomen that marks the point of former attachment of the umbilical cord.

Nitrogen Fixing Trees (NFTs)
Leguminous trees that take nitrogen from the air and fix it in the soil, making nitrogen available to other plants. They are often deep rooted, which allows access to nutrients in subsoil layers. The extensive root system stabilizes soil, while constantly growing and atrophying, adding organic matter to the soil while creating channels for aeration. There are many species of NFTs that provide useful products and functions, including food, wind protection, shade, animal fodder, fuel wood, living fence, and timber.

Offal
The entrails and internal organs of a butchered (slaughtered) animal.

Open
Referring to an adult female goat who has not yet conceived (become pregnant).

Organophosphate
A potent type of chemical insecticide and dewormer.

Ovulation
The release of a mature ovum (egg) from the ovary into the tube leading to the uterus of a female where it can then be fertilized by a sperm to develop into an embryo.

Palpate
To examine by touch, especially to score body condition and diagnose pregnancy, nutritional deficiencies or disease.

Parasite
An organism that gets its nutrition living in, with, or on another organism of another species, known as the host, usually to the detriment of the other organism.

Parturition
The act or process of giving birth.

Pasteurize
The partial sterilization of a substance and especially a liquid (as milk) at a high temperature and for a period of exposure that destroys objectionable organisms without major chemical alteration of the substance. To pasteurize milk, heat it to 63°C (145°F) and hold at this temperature for 30 minutes or "flash pasteurize" the milk by heating to 72°C (161°F) for 30 seconds.

Phenotype
The outward appearance of the animal due to its genotype and environment.

Placenta
The organ in the uterus of a female goat and other mammals that supplies the growing fetus with nutrients and oxygen through the umbilical cord. During birth, the

cord breaks and the placenta is expelled after the fetus. Goats and other ruminants have a "cotyledon" or "button type" placenta, that attaches to the uterus at many contact points, which looks like a button. Other mammals have different types of placentas.

Pluck
The heart, lungs and liver removed when an animal is slaughtered.

Pregnancy
The period of time from conception to birth during which a female carries a fetus in her uterus.

Profit
The difference between production costs and sale price for a product.

Protein
A complex, essential nitrogen-containing biological molecule. One of the building blocks of living things that makes up tissues and helps the animal resist disease. The structure of the body and enzymes for digestion are proteins.

Protein Supplement
Ingredients fed to goats, such as a cottonseed cake or commercial concentrate, that provide needed protein in a goat's diet.

Purebred Goat
A goat produced by mating a doe and a buck of the same breed.

Reproduction
The natural process by which new individuals are generated and the species perpetuated. The offspring share characteristics of both the mother and father.

Rumen
The largest of the four compartments of the ruminant's stomach, a fermentation vat where bacteria and protozoa break down fibrous plant material and other feedstuffs and synthesize essential proteins and vitamins.

Ruminants
Animals such as cattle, goats and sheep that have a complex stomach with four compartments and that ruminate and are able to digest cellulose.

Rut
A recurrent excitement that is observed in the male goat when he is ready to mate with the female.

Sanitary
Of or relating to health or the protection of health. Free from filth or pathogens that endanger health.

Semi-intensive Production
A husbandry system that uses a balance of grazing, concentrate feeding and housing as needed.

Silage
Green plant material that is chopped and stored without air so that it can ferment. It becomes very acidic and this allows it to be preserved for long periods, so that can be used as feed when forage is scarce.

Sire
A male parent.

Slaughtering Cradle
A wooden structure made to support the carcass of a goat while slaughtering.

Sperm
A male reproductive cell, which may fertilize the female ovum.

Spinous Process
The top part of the bone on each vertebra, which can be felt through the skin and flesh of the back and neck.

Sterile
Free from living organisms and especially microbes that might cause infection.

Subcutaneous Injection
Below or under the skin. Referring to an injection of a drug, or other substance under the skin.

Supportive Care
Care given in response to a need such as illness, kidding or environmental stress.

Sustainable Development
An activity that is socially, economically and environmentally responsible because it maintains or

restores available resources and is capable of continuing independently without outside resources.

Teat
An extended structure beneath the udder containing the canal that carries the milk from the mammary gland to the opening where it exits the body. Kids nurse milk from the teats. Farmers squeeze milk out of the animal through teats. Goats and sheep have two teats, while cows have four.

Teat Dip
A liquid sanitizing solution for disinfecting the goat's teats prior to and after milking.

Total Digestible Nutrients (TDN)
An old system to measure the available energy of feeds and the energy requirements of animals. TDN values are usually quoted as percentages for feeds and as amounts per day for the animal requirements. The values are usually calculated on feed analysis reports. The simplest and most commonly used formula for estimating is: TDN = Calories DE/0.044. One kilogram of TDN is equivalent to 44 calories of Digestible Energy.

Trauma
Severe physical or mental damage.

Udder (goat)
A bag-shaped structure underneath the abdomen containing the mammary gland, and a reservoir of milk, and characteristic of certain female mammals such as cows, sheep, and goats. In goats, the udder has two teats.

Udder Edema
Excessive swelling or redness of the udder.

Umbilical Cord
The cord containing blood vessels that attaches the fetus to the placenta of the mother and allows nutrients to pass from the mother to the fetus.

Uterus
A Y-shaped organ where the embryo is nurtured.

Vaccination
To administer a biological agent, usually by injection, that causes the animal to be immune to certain diseases.

Value Chain
The Value Chain is a diagram of all the "links" in a certain economic activity, like goat milk production.

Viscera
All of the guts or entrails of a goat after slaughtering.

Water Sac
The membrane enclosing the fetus (or fetuses) that is filled with amniotic fluid.

Wattles
Small natural appendages found on some goats, usually hanging from the neck, but may appear anywhere about the head and neck. They are also referred to as tassels.

Well-being
A good quality of life.

Wether
A castrated male goat.

Whey
The liquid left when milk proteins form curd in cheesemaking. It contains valuable minerals and can be used for drinking or cooking.

Withdrawal Time
An amount of time required following use of a medication in an animal before milk or meat can be safely used for human consumption.

Yearling
A goat 12 to 24 months old.

Zero Grazing
An intensive management system whereby animals are kept in a stall and all forage and water is provided by the caretaker. In this system it is very important that the animals get exercise and sunshine every day.

Zoonotic Disease
A disease communicable from animals to humans under natural conditions.

APPENDIX A – A SAMPLING OF GOAT BREEDS

Physical Characteristics and Environmental Suitability
There are about 500 breeds and types of goats worldwide. The vast majority of these are multipurpose animals, but usually one trait is predominant i.e., meat, milk, fiber production or survival under harsh conditions. There are certain breeds that are especially good for land reclamation.

The choice of a goat breed will depend on:

- Availability of breeding stock
- Size of farm
- Availability of housing for animals
- Family needs for meat, milk and income
- Local grazing and browsing rights
- The environment
- Culture
- Available markets
- Available labor
- Access to health care and supplemental feed

Anyone who has traveled in rural areas (or even some cities in the developing world) is aware of the large number of goats that browse the roadsides and pick at food wastes. These scavengers provide an important source of food and livelihoods for local populations. Local goats are often more acclimated to their environment and more disease-resistant.

Consult your local extension person for assistance in selecting the breed of goats best suited for a specific need or talk to a successful goat farmer in the area.

NON-EUROPEAN DAIRY BREEDS

Barbari

- Average lactation is 100 to 130 kg (220 to 290 lbs) in 200 to 250 days.

This small-sized breed is found mainly in northern India and Pakistan. They seem to thrive in intensive conditions and are generally stall-fed. The goats have short, erect 'tubular' ears and small pointed horns. The coat is often short, white hair with small tan or reddish-brown patches; or the entire coat is brown. It has a 'deer-like' appearance, with a small elegant head and somewhat prominent eyes. Males are bearded. The neck is long and slender, the shoulders are well developed and the legs straight and sturdy. The udder is well developed, with teats of medium length. Adult females weigh 27 to 36 kg (60 to 80 lb). The kidding interval is annual and twins and sometimes triplets are common.

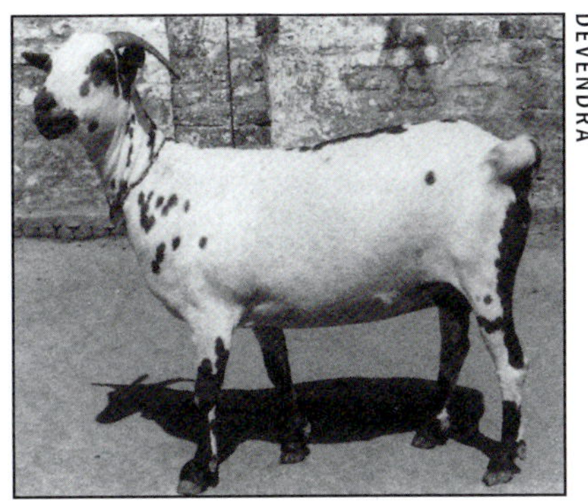
DEVENDRA

Beetal

- Average Lactation is 140 to 230 kg (310 to 510 lb) in 150 to 225 days. Maximum yield reported for a Beetal is 590 kg (1,300 lb) in 177 days.

This is another important breed, derived from the Jamnapari breed and one of the best milking breeds in India and Pakistan. The Beetal is a large spotted black goat with long, flat, curled and drooping ears, a Roman nose and wattles usually present on both sides of the lower neck. It is predominantly black, red or brown in color often with patches and spots of white. The body is compact and well-developed, with a dairy wedge-shaped conformation. The massive, broad head has a convex profile. Horns are quite long and slightly twisted in the male, and shorter in the female. The long legs are sturdy. The udder is well-developed, with long conical teats. The normal kidding interval is one year. Twins are common, with occasional triplets and even quadruplets. With its size advantage, the Beetal is in demand for its meat producing qualities: growth rates are high and the meat quality is excellent.

Damascus (Shami, Aleppo)

- Average lactation 500 to 550 kg (1,100 to 1,210 lb) in 260 days. The milk is comprised of 3.8 percent butterfat.

The Damascus is an outstanding dairy breed in the Middle East and is very prolific. Large, tall, roman-nosed, long lop ears, may be polled or horned. The horns may be sickle-shaped in the female, but twisted and spreading in the male. Coat is quite shaggy and colors are usually reddish brown, sometimes with white spots on the face, legs, and belly. Some are pied or grey. It is found in Cyprus, Syria, Iraq, Jordan, Israel and Lebanon. Milk production is usually highest during the sixth lactation. Average litter size at weaning is 1.68 kids; if properly fed, can be bred at 8-10 months of age.

GEORGE F.W. HAENLEIN

Jamnapari

- Average lactation is 60 to 200 kg (130 to 440 lb) over 210 to 240 days. Maximum milk yield is 540 kg (1,200 lb) in 250 days.

Originating from Northern India, the Jamnapari is a dual purpose goat that has undergone some selection for milk production and is one of the most popular dairy goats in India. It is one of the largest breeds and has been used extensively to upgrade local goats (in neighboring countries as well). There is a considerable variety in the colors of its coat but often is white with patches of tan or black on the head. It is a tall, rangy, slender goat and is normally horned (upright

GEORGE F.W. HAENLEIN

and often twisted horns). It has a large convex face and large pendulous long ears. The breed has some resemblance to the Nubian except for longer hair on the rear legs. The udder is round and very well developed, often hidden from view at the rear by thick, hairy 'feathering' on the buttocks. Twins are quite common and occasionally triplets, though most have single kids. They feed on tree-loppings, crop stubble and Acacia pods.

Nigerian Dwarf

- Average lactation is 250 to 300 kg (550 to 660 lb) over 300 days. Milk has six to 10 percent butterfat.

They have a higher protein content than most other dairy goat breeds. Dwarf goats can breed year around. For the most part, Nigerian Dwarfs are a hearty breed and seldom have kidding problems. New babies are about two pounds at birth and grow quickly. Does can be bred at seven to eight months of age. Dwarfs can have several kids at a time, with triplets and quads being common. Dwarfs are generally excellent mothers. Dwarf goats are resistant to Trypanosomiasis. They are fine-boned, with a long, refined neck and straight or slightly dished profile. Ears are erect and both sexes are horned. The hair is short and fine with a wide color range including brown, black and golden, often with random white markings.

CORNERSTONE FARM

Zaraibi (Egyptian Nubian, Theban)

- Average lactation is 250 to 300 kg (551 to 661 lb) over six-month cycle. A good producer will give up to 500 kg (1,102 lb).

The name translates into "barn type", indicating that they were managed in confinement or tethered around dwellings. The Zaraibi is the rarest of the three ancestors of the Anglo-Nubian and is found in decreasing numbers in Egypt and Sudan. It is a medium-sized long-legged goat. The most prominent feature is the Roman nose. It's coloring is cream, red or light brown with dark spots or black with spots. It is more often polled than horned. It has long legs, a long and deep body, and wide, long drooping ears that have a fold at the tip, especially in young kids. The face is convex, with often undershot jaw. The Zaraibi from sub-Saharan Africa is also tall and rangy, but deeper bodied. Average litter size is 1.5 kids.

GEORGE F.W. HAENLEIN

EUROPEAN/SWISS DAIRY BREEDS

European Breeds were introduced to the United States and other areas as the quickest means for increasing milk production. These breeds are generally above average in size with mature does weighing 45 to 55 kg (100 to 120 lb) and bucks 60 to 75 kg (130 to 165 lb). When good nutritional standards are met, milk production ranges from 350 to 900 kg (770 to 1,980 lb) in a 300 day lactation. Nubian, Saanen, Alpine and Toggenburg are featured in this course. They can be considered both as purebred producers and for their potential in crossing with local goats.

Alpine

TATIANA STANTON

- Average lactation is 600 to 900 kg (1,320 to 2,000 lb) in 250 to 305 days.

The Alpine is a highly developed milk breed and originated in the Swiss and Austrian Alps. They are medium to large in size and are hardy and adaptable animals, thriving in many climates. The face is either straight or slightly dished with a wide forehead, light brown eyes, a gentle expression, erect 'horn shaped' ears of medium length, long body and well developed stomach. The coat is short, except for a fringe of longer hair along the back and on the thighs; the males' coat is much longer than the females, and closer, covering much of the forequarters. The color patterns can vary with different shades in the same animal; strongly marked by streaks or washes in all the tones of black, grey, chestnut, fawn, and even brown mauve; white neck, gray markings on head, off white front quarters, black front quarters, black with white such as belly and facial trim, spotted or mottled. Due to many years of intensive selection, the breed has excellent dairy conformation and good fertility. It is a quick-growing goat, reaching about 0.9 m (35 inches) in the female and up to 1 m (40 inches) in the male and weighing up to 90 kg (200 lb).

Nubian/Anglo Nubian

TATIANA STANTON

- Average lactation is 700 to 900 kg (1,540 to 1,980 lb) in 275 to 300 days, with 5 percent butterfat.

The Nubian type goat originated in East Africa. Distinctive characteristics show the ancestry of Indian Jamnapari, Indian Chitral, British native and Zaraibi. The Nubian is a relatively large, graceful dairy goat. The distinguishing characteristic is the head. The profile between the eyes and the muzzle is strongly convex, creating its Roman nose. The head is short with no tassels. The ears are long, wide and pendulous. They lie close to the head at the temple and gently flare out and forward. Nubians generally have long bodies and long legs. The hair is fine, short and glossy. The colors vary greatly, with black, tan or red colors predominating, often spotted or dappled. Nubians often adjust more readily to extreme heat. Anglo-Nubians have high potential for milk production and prolificacy. The Nubian is used in the Middle East, South America and the Caribbean to improve the milk and meat yield of native goats.

Saanen

- Average lactation is 800 to 900 kg (1,760 to 1,980 lb) in 275 to 300 days.

OKLAHOMA STATE UNIVERSITY

The Saanen is a large animal with strong bones. The breed originated in Switzerland and is the largest Swiss breed averaging 50-75 kg (110 to 165 lb) in body weight. It is an all white or creamy-colored animal. The face is straight, the ears are upright and alert. The hair is short and fine. Wattles may or may not be present. Males sometimes have a longer fringe of hair along the back and on the legs, often with a 'skirt' or 'saddle' of longer hair and a beard. Saanens are vigorous animals, well-suited to intensive systems, and renowned for high milk production and excellent udders. However, Saanens do not fare well in strong tropical sunshine, where it must have access to shade at all times. Milk yields are greatly reduced in tropical areas. The Saanen has been introduced into all continents and throughout the tropics and is the most widely distributed of the improved breeds.

Toggenburg

- Average Lactation is 600 to 900 kg (1,320 to 1,980 lb) in 275 to 305 days.

NESHAMINY ACRES, CHALFONT, PA

The Toggenburg is a Swiss breed from the Toggenburg Valley in Switzerland. It has been called the oldest breed of Switzerland and was one of the first to be imported to the United States. It is medium-sized and sturdy in appearance. The hair is usually short to medium and is fine and soft; however, some individuals have a heavy hair coat. The color varies from mid-brown to fawn, mouse-grey and almost silver. The distinct white markings are two white stripes down the face from above each eye to the muzzle; hind legs white from hocks to hooves; forelegs white from knees downward with dark below the knee acceptable; a white triangle on either side of the tail; a white spot may be at root of wattles or in that area if no wattles are present. The ears are erect and carried forward. The nose is straight or slightly dished. It may be horned or polled, and with or without wattles.

MEAT BREEDS

Barki

JOHANNA STIGTER

Also called Arabian or Bedouin goat, is the desert goat of Somalia, Syria, Israel, Egypt and Jordan. Barki are used by pastoralists for meat, milk, hair, and gifts. They have course, long black hair, long, twisted horns which are scimitar-shaped and long lop ears that are carried laterally. More or less straight profile, though slightly arched. Adult body weights are about 15 to 22 kg (33 to 50 lb). They are well adapted to the harsh conditions of the desert, and can be watered once every two to four days. They can produce an average milk yield of 2 kg (4.0 lb) per day when feed is adequate. Twins are common. Most are black, often with white spots on head and legs. Males and most females are horned. Ears are medium length. Average 1.2 kids per litter.

Boer

CHARLEY CHAMBERLIN

"Also known as Afrikaner, South African Common Goat, the Boer is an improved indigenous breed with some infusion of European. The name is derived from the Dutch word "Boer" meaning farmer. The Boer goat is primarily a meat goat full of muscle and vigor. It is a horned breed with lop ears and showing a variety of color patterns and a convex profile. The coat is soft, smooth and glossy, and the hair is short to medium length with no woolly undercoat. The head is strong, with a roman nose and large eyes. The prominent, rounded horns are dark and set well apart, growing with a gradual backward curve. The broad drooping ears are of medium length. The shoulders are broad and fleshy, the chest is broad, the ribs well sprung, the brisket deep and broad, the withers broad and well filled. In females, the udder and teats are well developed. It is a docile breed and easy to manage, producing average weaning rates in excess of 160 percent. The mature Boer Goat buck weighs between 110 to 135 kg (240 to 300 lb) and does between 90 and 100 kg (200 to 220 lb). A kidding rate of 200 percent is common for this breed. The Boer goat also has an extended breeding season making possible three kiddings every two years.

Nanjiang Yellow

HEIFER INTERNATIONAL CHINA

The Nanjiang Yellow goat was developed in China between 1960 and 1990 as a Sichuan meat goat from Chengdu Brown and Nubian males crossed with local goats. The Nanjiang Yellow goat is the one of the best goat breeds for meat in China. It was bred up successfully using four breeds and over 40 years in the mountainous areas in and around Nanjiang County, Sichuan Province, China. Its hair is short and is brown and drab; the ears are mid-sized and erect. The average birth weight is 2.2 kg (5.0 lb) for male and 2.0 kg (4.4 lb) for female. The two months weaning weight is 11 kg (24 lb) and 10 kg (22 lb); the 12 months weight is 38 kg (80 lb) and 31 kg (68 lb); the adult weight is 67 kg (150 lb) and 46 kg (101 lb) respectively. Dressing weight at slaughter at 12 months is about 50 percent.

Spanish

Originally from Spain, the Spanish goat came to the United States via Mexico. The Spanish goat has the ability to breed out of season and is an excellent range animal because of its small udder and teats. In addition, Spanish goats are usually characterized as being very hardy, able to survive and thrive under adverse conditions with only limited management inputs. Within the general group of "Spanish goats" there are those that are purely Spanish, whereas others represent a mixture of all genotypes introduced to the area. There have been obvious infusions of dairy and Angora blood in many Spanish herds but no organized attempt has ever been made to use them for milk or mohair production. Until recently, these goats were kept mainly for clearing brush and other undesirable plant species from pasture lands. In recent years, the escalating demand for goat meat and the expanding interest in cashmere production have focused attention on the Spanish goat.

FIASCO FARM

West African Dwarf

UMARU SULE

The West African Dwarf goat is about 40 to 50 cm (16" to 20") in height and 20 to 30 kg (40 to 66 lb) in weight. It has disproportionately short legs, a short, broad head and plump body. The ears are short to medium-sized and erect or carried horizontally, and the horns are very short (wide-based in the male, more slender in the female); a few are polled. The coat is fairly short, but the male often has a thick, longer fringe along the back, and a beard. Colors vary from fawn to brown with black back-line, belly and tail, or black, brown, grey and white or tricolored, including white and yellow. It has early sexual maturity and can be very fertile and prolific, often producing twins and occasionally triplets and quadruplets. Growth rate and milk yield are very low, but they will breed at all times of the year. They are used almost exclusively for meat production. One of the most important characteristics is that they are resistant to the disease Trypanosomiasis which is spread by the Tsetse fly and they can thrive without extensive protection measures.

MULTIPURPOSE BREEDS

Most goats could be classified as multipurpose, providing milk and or meat, skins, manure etc. The following goats can be classified as multipurpose.

Baladi ("rustic" in Arabic)

This is the native goat of the Middle East, and is found from Morocco to Syria. They are prolific and breed in any season. There is a high demand for them as meat and they also produce milk and hair. They are small slender goats with a straight or slightly convex facial profile. They probably originated in the Nile valley and Delta and their appearance can be varied, though usually with long-hair and a relatively small head, straight profile and drooping ears of medium to large size. Wattles are common, but only bucks have beards. The males have small horns that curve backwards and then downwards; most females are polled. Mature animals weigh 30 to 35 kg (66 to 80 lb). They are prolific, having twins and triplets. The coat is usually black but might also be white, red, grey or a mixture of colors, including pied.

BANGLADESH AGRIC. UNIV.

Black Bengal
The Black Bengal has a number of outstanding features. They are widely distributed in Bengal, India and in the northern part of East Pakistan. The breed is economically important because in addition to its production of meat of high quality, it also produces a skin for which there is considerable demand. The skin is used extensively for making high quality shoes.

AN PEISCHEL

Kiko
The Kiko meat goat from the South Island of New Zealand has the genetic heritability of foraging. The Kiko was raised under varying climatic conditions and on rugged terrain. The main selection characteristics for the breed are survival in rugged hill country and growth rate of kids under poor nutritional conditions and pasture kidding. The first Kiko genetics arrived in the United States in the early 1990's. Today the Kiko goat is large framed and early maturing. The primary characteristic of the Kiko is its hardiness, ability to achieve substantial weight gains under natural browsing and foraging conditions without supplemental feeding. Kiko kids grow rapidly, averaging 32 kg to 41 kg (70 to 90 lb) at eight months.

While Kikos may come in every color, the white coat is a dominant gene. Does rarely require help during kidding and most kidding is under foraging conditions. Although not bred for milk production Kikos can be used as dairy animals and after weaning twins give up to 3 liters per day during a three month lactation period without supplemental feeding.

Fiber and Skins
Cashmere, Angora, Pashmina, Red Sokoto are examples of goats used for fiber and skin. While this book does not deal extensively with the production of fiber, these goats play an important role in the economies of farmers. Most goat farmers will find a use for goat skins to make drums, clothing or other specialty items.

DAIRY DOE RECORD SHEET

Breed:	Reg #:	Born:
Name:	Tattoo #:	Disbudded:
Color:		

Left

Right

Sire:	Reg #:
Dam:	Reg #:

Sire:	Reg #:
Dam:	Reg #:
Sire:	Reg #:
Dam:	Reg #:

LACTATIONS

Freshening Age	Total Days	Lb. Milk	Butterfat	%

ORIGIN

Farm Reared:
Purchased From:
Date:
Cost:
Date Died:

KIDDING					
Bred	Due	Kidded	No. in Litter	Sire	Comments

HEALTH		
Date	Condition	Treatment

INDIVIDUAL DAILY MILK RECORD SHEET

Name of Goat:________________ No:________________ Year:________________

Weigh milk and record in kilos or lbs. If no scale, measure after straining and record in liters or quarts.

DAY	JAN		FEB		MAR		APR		MAY		JUN		JUL		AUG		SEP		OCT		NOV		DEC		TOTAL
	am	pm	am	pm	am	pm	am	pm	am	pm	am	pm	am	pm	am	pm	am	pm	am	pm	am	pm	am	pm	
1																									
2																									
3																									
4																									
5																									
6																									
7																									
8																									
9																									
10																									
11																									
12																									
13																									
14																									
15																									
16																									
17																									
18																									
19																									
20																									
21																									
22																									
23																									
24																									
25																									
26																									
27																									
28																									
29																									
30																									
31																									
TOTAL																									

INDIVIDUAL DAILY MILK RECORD SHEET

Weigh milk and record in kilos or lbs.
If no scale, measure after straining
and record in liters or quarts.

Name of Goat:______________________ No:______________ Year:______________

DAY	JAN		FEB		MAR		APR		MAY		JUN		JUL		AUG		SEP		OCT		NOV		DEC		TOTAL
	am	pm	am	pm	am	pm	am	pm	am	pm	am	pm	am	pm	am	pm	am	pm	am	pm	am	pm	am	pm	
1																									
2																									
3																									
4																									
5																									
6																									
7																									
8																									
9																									
10																									
11																									
12																									
13																									
14																									
15																									
16																									
17																									
18																									
19																									
20																									
21																									
22																									
23																									
24																									
25																									
26																									
27																									
28																									
29																									
30																									
31																									
TOTAL																									

MEAT DOE RECORD SHEET

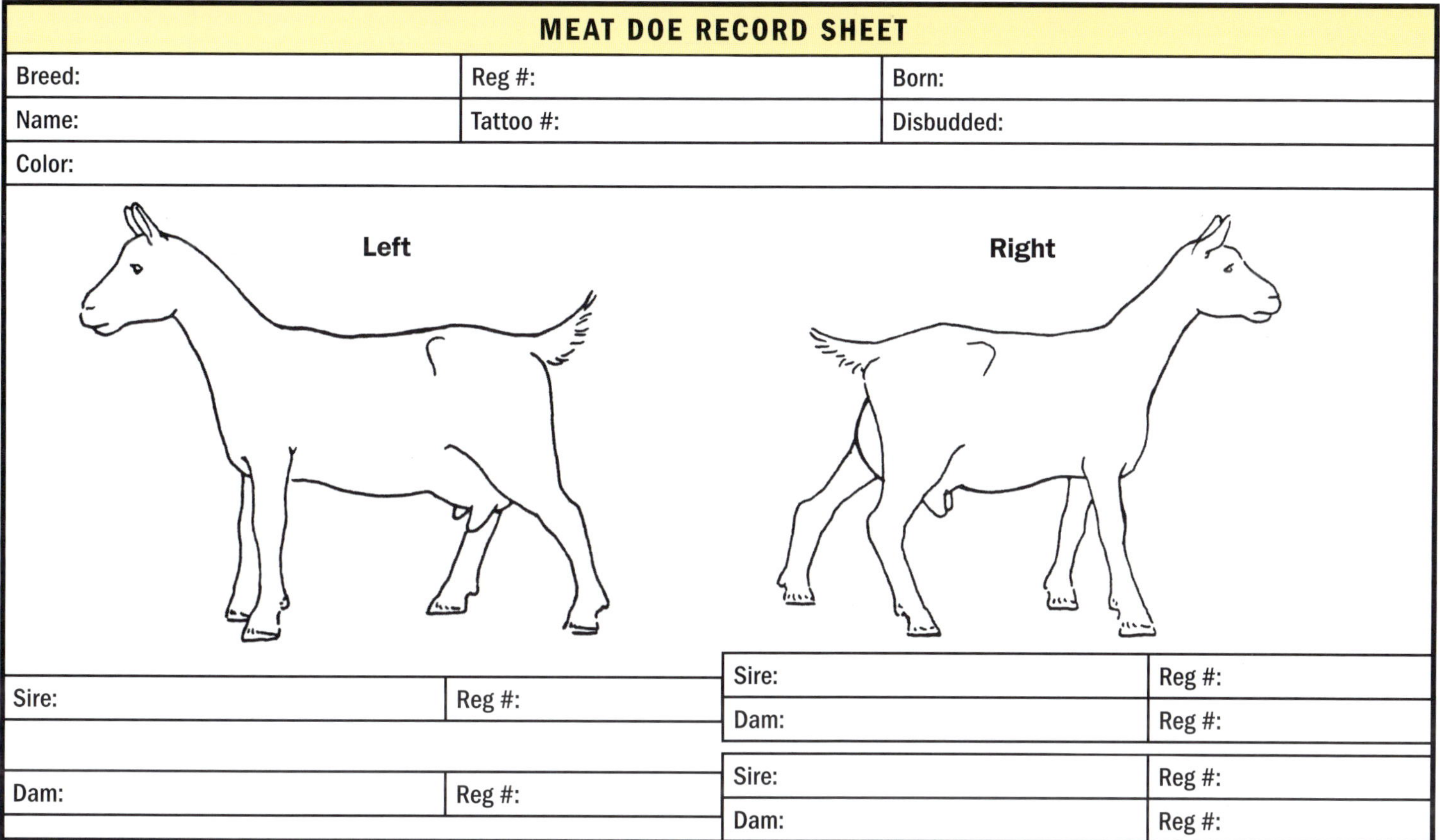

MEAT DOE RECORD SHEET

Breed:	Reg #:	Born:
Name:	Tattoo #:	Disbudded:
Color:		

Sire:	Reg #:
Dam:	Reg #:

Sire:	Reg #:
Dam:	Reg #:
Sire:	Reg #:
Dam:	Reg #:

Date	Body Condition Score	FAMCHA Score

ORIGIN
Farm Reared:
Purchased From:
Date:
Cost:
Date Died:

BODY CONDITION SCORE	
Thin	1-3
Moderate	4-6
Fat	7-9

KIDDING

Date Bred	Sire ID	Date Due	Date Kidded	Kid ID	Sex	Birth Weight	Weaning Weight

HEALTH

Date	Condition	Treatment

BUCK RECORD SHEET

Breed:	Reg #:	Born:
Name:	Tattoo #:	Disbudded:
Color:		

Sire:	Reg #:	Sire:	Reg #:
		Dam:	Reg #:
Dam:	Reg #:	Sire:	Reg #:
		Dam:	Reg #:

ORIGIN

Farm Reared:
Purchased From:
Date:
Cost:
Date Died:

BREEDINGS

Date	Doe

BREEDINGS

Date	Doe

HEALTH

Date	Condition	Treatment

APPENDIX C – BUDGETING

BUDGETING

First list the overall goals for the goat enterprise. Then use the chart to list your resources, where you will get them and at what cost. The second column is for benefits, such as the price you can sell each liter of milk. The value for the household is the cost that milk consumed would have cost, plus other "intangible" benefits such as healthier children from better nutrition. Include all benefits from the goats, including manure and sale of kids for meat.

YEARLY FARM BUDGET FOR GOAT ENTERPRISE

Item	Where Available	At What Cost (Expense)	For What Benefit	Value for Household (Income)
Goats How many?				
Zero grazing unit and fencing				
Forages				
Concentrates and mineral salt				
Milking equipment				
Health care				
Sale of milk and milk products				
Sale of meat and meat products				
Sale of kids				
Exchange of milk and manure for labor				
Manure for garden				
Milk and meat for family				
Hired labor				

YEARLY FARM BUDGET FOR GOAT ENTERPRISE				
Item	Where Available	At What Cost (Expense)	For What Benefit	Value for Household (Income)

APPENDIX D – CONVERSION TABLE

TEMPERATURE		
To convert Celsius (°C) to Fahrenheit (°F)		
36.0°C	=	96.8°F
39.8°C	=	103.6°F
36.5°C	=	97.7°F
37.0°C	=	98.6°F
37.5°C	=	99.5°F
38.0°C	=	100.4°F
38.2°C	=	100.8°F
38.4°C	=	101.1°F
38.6°C	=	101.5°F
38.8°C	=	101.8°F
39.0°C	=	102.2°F
39.2°C	=	102.6°F
39.4°C	=	102.9°F
39.6°C	=	103.3°F
40.0°C	=	104.0°F
40.2°C	=	104.4°F
40.4°C	=	104.7°F
40.6°C	=	105.1°F
40.8°C	=	105.4°F
41.0°C	=	105.8°F
41.2°C	=	106.2°F
41.4°C	=	106.5°F
41.6°C	=	106.8°F
41.8°C	=	107.1°F
42.0°C	=	107.5°F
Note: To reduce degrees F to degrees C, subtract 32, then multiply by 5/9. To reduce degrees C to degrees F, multiply by 9/5 then add 32.		

1 milliter (mL)	=	1000 microliters
1 liter (l)	=	1000 mL

To convert liters (L) to ounces (oz), pints, quarts (qt), and gallons (gal)		
¼ L	=	8 ½ oz
½ L	=	1 pt 1 oz
1 L	=	1 qt 2 oz
4 L	=	1 gal 7 oz
10 L	=	2 ¾ gal
25 L	=	6 ¾ gal
50 L	=	13 ¾ gal
100 L	=	26 ½ gal

To convert ounces (oz), pints (pt), quarts (qt), and gallons (gal) to milliters (mL) and liters (L)		
29. 6 mL	=	1 oz
59.2 mL	=	2 oz
88.8 mL	=	3 oz
118.4 mL	=	4 oz
148 mL	=	5 oz
177.5 mL	=	6 oz
207 mL	=	7 oz
236.6 mL	=	8 oz
473.2 mL	=	1 pt
946 mL (or 0.946 L)	=	1 qt
1.893 L	=	2 qt
2.830 L	=	3 qt
3.785 L	=	4 qt

CONVERSION TABLE

WEIGHTS		
1 gamma	=	1 microgram (mcg)
1,000 mcg	=	1 milligram (mg)
1,000 mg	=	1 gram (g)
1,000 gm	=	1 kilogram (kg)
1, 000 gm	=	1 kg = 2.2 lb

To convert ounces (oz) and pounds (lb) to grams (gm) and kilograms (kg)		
28.3 g	=	1 oz
56.7 g	=	2 oz
85.0 g	=	3 oz
113.3 g	=	4 oz
141.6 g	=	5 oz
170.0 g	=	6 oz
198.3 g	=	7 oz
226.6 g	=	8 oz
255 g	=	9 oz
283.5 g	=	10 oz
311.8 g	=	11 oz
340 g	=	12 oz
368.3 g	=	13 oz
396.6 g	=	14 oz
425 g	=	15 oz
453.6 g	=	16 oz (1 lb)
453.6 g	=	1 lb
907.2 g	=	2 lb
1.4 kg	=	3 lb
1.8 kg	=	4 lb
2.3 kg	=	5 lb
4.5 kg	=	10 lb
22.7 kg	=	50 lb
45.4 kg	=	100 lb

LINEAR MEASURE		
1 millimeter (mm)	=	0.04 inch (in)
1 centimeter (cm)	=	0.4 in
1 decimeter	=	4 in
2.54 cm	=	1 in
30.48 cm	=	1 foot (ft)
91.44 cm	=	1 yard (yd)
1 cm	=	10 mm
100 cm	=	1 meter (m)
1 kilometer (km)	=	1000 m

To convert ounces millimeter (mm) and centimeter (cm) to inches (in)		
3 mm	=	⅛ in
6 mm	=	¼ in
11 mm	=	½ in
18 mm	=	⅔ in
2.5 cm	=	1 in
5.0 cm	=	2 in
7.5 cm	=	3 in
10 cm	=	4 in
12.5 cm	=	5 in
15.0 cm	=	6 in
17.5 cm	=	7 in
20.0 cm	=	8 in
22.5 cm	=	9 in
25.0 cm	=	10 in
27.5 cm	=	11 in
30.0 cm	=	12 in

APPENDIX E – GOAT AND GOAT RELATED ORGANIZATIONS

American Association of Small Ruminant Practitioners (AASRP)
2413 Nashville Rd
Suite 112; MS-C13
Bowling Green, KY 42101 USA
www.aasrp.org

American Boer Goat Association
1207 S. Bryant Blvd, Suite C
San Angelo, TX 76903 USA
Phone: (325) 486-2242
Fax: (325) 486-2637
E-mail: info@abga.org
www.abga.org

American Dairy Goat Association
209 West Main Street - PO Box 865
Spindale, NC 28160 USA
Phone: (828) 286-3801
Fax: (828) 287-0476
E-mail: info@adga.org
www.adga.org

American Livestock Breeds Conservancy
PO Box 477
Pittsboro, NC 27312 USA
Phone: (919) 542-5704
www.albc-usa.org

American Meat Goat Association
PO Box 676
Sonora, TX 76950 USA
Phone: (915) 835-2605
www.meatgoats.com

Domestic Animal Diversity Information System DAD-IS
Food and Agricultural Organization of the United Nations
www.fao.org/dad-is

E. (Kika) de la Garza Institute for Goat Research
Langston University
Agricultural Research and Extension Programs
P.O. Box 730, Langston, OK 73050 USA
Phone: (405) 466-3836
www2.luresext.edu/goats/index.htm

FARM-Africa
Clifford's Inn
Fetter Lane
London
EC4A IBZ
UK
Phone: 44 (0) 20-7430-0440
E-mail: farmafrica@farmafrica.org.uk
www.farmafrica.org.uk

Heifer International
1 World Avenue
Little Rock, AR 72202 USA
Phone: (501) 907-2600
www.heifer.org

International Goat Association
1 World Avenue
Little Rock, AR 72202 USA
Phone: (501) 454-1641
E-mail: goats@heifer.org
www.iga-goatworld.org

International Livestock Research Institute (ILRI)
PO Box 30709
Nairobi, 00100 Kenya, Africa
E-mail: ILRI-kenya@cgiar.org
www.ilri.org

Oxfam International
Oxfam House
John Smith Drive
Cowley
Oxford
OX4 2JY
UK
www.oxfam.org

Wool and Wattles
AASRP Newsletter
Cornell University
Ithaca, NY 14853 USA
www.aasrp.org

World Neighbors
4127 NW 122nd Street
Oklahoma City, OK 73120 USA
E-mail: info@wn.org
www.wn.org

These recipes for milk and meat products have been gathered from personal experience, from friends and from published materials. When preparing any food product, do it in a cool, clean environment. Always wash your hands thoroughly with soap and water and rinse before handling food.

MILK PRODUCTS

Cheese and Yogurt Making: Getting Started

Always use good quality milk for cheese making. Rancid, off-flavored or mastitic milk or milk contaminated with antibiotics should be avoided. Cheese is milk protein (casein) and fat coagulated into curds by the addition of acid. Low fat cottage, Romano and Parmesan cheeses are traditionally made with skim milk, but most other cheese are derived from whole milk.

The acid can be provided by adding vinegar, lemon juice or citric acid to milk or by the inoculation of milk with a lactose (milk sugar) consuming bacteria, which digests the lactose and produces lactic acid, thus acidifying the milk. If planning to inoculate milk with a bacterial culture to produce cheese, pasteurize the milk first to kill off existing bacteria that may compete with the culture being introduced. To pasteurize milk, heat it to 63°C (145°F) and hold at this temperature for 30 minutes or "flash pasteurize" the milk by heating to 72°C (161°F) for 30 seconds. Quickly cool milk immediately after pasteurization. This can be done by placing the container in cool water.

Rennet is used to form a firm, uniform curd. It is especially important in making hard cheeses. Plant rennet is derived from a mold, *Mucor meihei*. Animal rennet is found in the lining of the true stomach of suckling calves and goat kids. Both rennets can be purchased in tablet or liquid form and need to be diluted in cold water before adding to milk. A small amount of rennet goes a long way and your culture, once developed, can be saved from a previous batch.

Equipment

It is best to use stainless steel, glass or enamel containers when making cheese and yogurt. This is because aluminum and cast iron equipment may react with the acid or release metallic acids into the milk, causing variations in taste and coloring.

It is also important to have a fine mesh cheesecloth or other cloth to strain the whey from the curd. Although cheese molds can be bought commercially, a mold can be made out of any round container with holes in it or a coarsely woven cloth. Any mold should hold at least one-half kilo of curd.

For certain pressed cheese, a weight may be needed to remove excess whey or water. Make a press out of a jar full of water, a brick wrapped in a clean cloth or any other weight that will fit on top of the mold. The weight should be 1 to 2 kg depending on the size of the cheese.

All equipment should be washed and sterilized before each use. A sterilizing solution can be made by adding one tablespoon of bleach to a gallon of water. Cheesecloths and other cloths can be sterilized by boiling in water before use.

Basic Equipment

ROSALEE SINN

- Stainless steel pan
- Large spoon
- Small and large strainer (sieve)
- Cheesecloth
- A sharp cutting knife
- A small knife
- Lemons, vinegar or limes
- Cooking thermometer
- Measuring cup
- Liquid rennet or tablets
- Yogurt
- Salt and black pepper
- Herbs
- Clean, fresh water
- A fire or stove
- Small cheese molds (bottom right of photo)
- Two large stainless steel or glass bowls (2 to 4 liters or ½ to 1 gallon)
- A clean, fly-proof place to hang and drain cheese

Fresh Cheese

This is a wonderful way to begin. The cheese is sweet tasting, smooth and delicious. It takes about 1 hour.

ROSALEE SINN

Ingredients

- 2 liters (about 8 cups) goat's milk
- ½ cup of lemon juice or vinegar
- Salt
- Black pepper
- Herbs

Instructions

Chop herbs and set aside. Squeeze ½ cup of lemon juice or measure ½ cup of vinegar.

ROSALEE SINN

Place a double layer of cheese cloth or a single layer of cloth with a large weave in the sieve. Place sieve in the large bowl.

Pour goat's milk in pan and place over medium heat. Bring milk to just under a boil, stirring so it will not stick or burn. Remove from heat, keep stirring and add ½ cup lemon juice or vinegar. The acidity of the vinegar or lemon juice will determine how much to add. If curds do not form add a little more. Strain curds at once through a double layer of cheese cloth or a cloth with an open weave. (In the photo the herbs were placed in the sieve before the curds were poured in, but they can be added later.)

ROSALEE SINN

Lift the cloth and tie with a string. Hang over the bowl for about 15 minutes until the whey has all drained into the bowl. Give a little squeeze to reduce liquid. Pour the whey into a container to save for other cooking or drinking. Empty the cheese into a bowl, scraping the cheese cloth with a knife so as not to lose any. You will have about ½ liter of cheese for each 2 liters of milk.

ROSALEE SINN

Mix with a spoon. Add a pinch of salt, black pepper to taste and finely chopped herbs. (dill, rosemary, basil, thyme, sage, or chives all work well and can be mixed together.) Stir salt, pepper and herbs into the cheese and form a flat circular disk about 5 cm (2 inches high.) Pat with spoon until firm and well shaped. Cover with damp cheese cloth or plastic wrap. Put in a cool place. Enjoy with bread, crackers or slices or fruit.

ROSALEE SINN

Use the whey (left after the curds are drained) for cooking or drinking. Whey contains valuable vitamins and minerals and can be used instead of other liquids when making bread or cooking beans, corn or potatoes to add extra nutrition.

Neshaminy Acres Goat Cheese

Equipment

- Clean stainless steel utensils
- 2 stainless steel pans that will hold 6 liters (1½ gal) of milk (1 for heating milk; 1 for draining)
- Strainer or colander
- Fine cheesecloth
- Clean 1.1 kg weight (½ lb) (for harder cheese)

Instructions

Pasteurize (heat) milk at 74°C (165°F) for 15 seconds. Place in cold water in sink or larger pan to cool. Change water once or twice or add ice to speed cooling. When cooled to less than 30°C (86°F) add Fromage Blanc Direct Set Culture from New England Cheesemaking Supply Co. Add culture at the rate of 1 package to 4 to 5 liters of milk (package says 1 gallon). After stirring in the culture cover pan with a clean dishtowel and set in a warm place for 12 to 16 hours or until whey appears on top (sometimes there's a lot of whey, sometimes very little.) Insert a knife in the center of the cheese and it should hold the slit and a bit of whey should be visible. Let it set for up to 24 hours. Early lactation milk is a bit slower to firm up.

Line a colander/strainer with cheesecloth (the "butter or chef's cheesecloth is better than the gauze type from supermarket, but either will work) and put this over a pan which will catch the whey. Use a long knife and cut the curds into ½ inch squares. Gently ladle the cubes into the cheesecloth-lined strainer and allow to drain. The curd should be above the liquid whey. Use a clean spatula to scrape the cheese from the sides of the cheesecloth. Leave this at room temperature until the majority of the whey is off. This may take two to three hours. Place in refrigerator or a very cool place and let drain for 24 or more hours. For firm, soft cheese, add some weight on top.

After draining, remove from cheesecloth and add any of the following to taste: salt, pepper, herbs i.e., finely chopped parsley, chives, dill seed, rosemary, onion salt, garlic salt or any combination. For plain cheese, simply add salt to taste. Plain cheese can be rolled in black pepper or finely chopped or dried herbs. Dried cranberries, diced dates and dried apricots, walnuts or pecans can also be added.

Note: Fromage Blanc Direct Set Culture can be obtained from www.cheesemaking.com. Alternatively you could use any mesophilic culture and add rennet.

Pressed Cheese
Use 7.5 liters (2 gal) of milk. Heat to 63°C (145°F). Cool to lukewarm immediately by setting in cold water. Add ¼ cup yogurt or buttermilk. Stir well. Dissolve ⅛ of a rennet tablet in ¼ cup cold water or use several drops of liquid rennet. Add to milk and stir well. Let set until firm or when you insert a knife it comes out clean.

Cut curd in ¼ inch strips up and down and crosswise until you have small pieces. Let stand until all the whey has separated from the curd—at least ½ hour to one hour. Stir, strain through a cheesecloth or other similar cloth. Squeeze out as much moisture as possible and press overnight. A jar filled with water or a brick wrapped in a clean cloth can be placed on top of cheese to press out water. Keep in cool place.

Semi-hard Goat Cheese
(Takes 2 hours) Into an 8 liter pot (2 gallons), pour 6 liters (or 1.5 gallons) of fresh-drawn goat milk. (Do not cool milk - this gives cheese a rubbery texture.) Add 1 cup buttermilk or yogurt and stir. Dissolve ½ of a rennet tablet in ¼ cup of cold water, crush with spoon. Add rennet solution to warm milk, stir for one minute.

Let stand in a warm place until a firm curd forms, about 45 minutes. Use a long knife to cut curd into cubes two ways, vertically and at an angle. Warm slowly, about 20 minutes (very low heat) to about 42°C (110°F), stirring frequently with a spoon to keep curds from sticking together.

Pour curds and whey into a cloth-lined strainer. Drain curds. Add 1 tsp salt and mix with hands. Add herbs if desired. Squeeze out whey by twisting cheese cloth. When firm remove from cloth and wrap in plastic. Ready to eat.

Esther Kinsey's Feta Cheese
Feta Cheese is good in the tropics because it can be stored easily in salt brine without refrigeration for several months. There is also a good market for feta.

Make semi-hard cheese (above recipe). Let stand for 12 hours in a cheese cloth to drain. Cut into wedges of not more than 10 cm (about 4 inches) of any dimension. Rub salt on all surfaces. Let it sit again in a cool place for 6-12 hours before cutting and putting into the brine bucket.

The mixing of salt brine is an art. Use 750 grams (1.6 lb) of salt per 10 liters of water (a liter is 33.8 ounces; a quart is 32 ounces). Keep stirring and add more salt. When a raw egg will float on the surface exposing about 2 cm (about 1 inch), it is the right density. Keep the wedges submerged in the brine in a bucket using a plate of same diameter as the bucket. Check now and then to remove any unsightly scum on the surface. In six weeks the feta cheese will be ready. To reduce the saltiness of the feta, you may soak it for two hours in milk.

Ricotta Cheese

Surplus whey from cheesemaking can be used to make ricotta cheese. It is easy! Take 2 liters (½ gal) of whey and heat it until a creamy substance rises to the top. Pour in three cups of milk and heat it until it is scalding. Then stir in 3½ tablespoons of vinegar or lemon juice. Stir briskly until the curd floats, keeping it very hot, not boiling. Remove the curd with a large spoon as it rises to the top. Place in a strainer lined with a cheesecloth. Drain it for 7 to 8 hours at room temperature. Salt to taste and place in cool place. Ricotta can be used with a number of dishes, including lasagna.

Brousse Cheese

Source from New England Cheesemaking Supply Co., Ashfield, MA 01330

Heat one gallon of goat's milk to 29°C (85°F) in a sterilized, stainless steel or enamel pan immersed in a larger pan of water. Stir in thoroughly 1 tablespoon of cheese starter culture (or 1 tablespoon of either buttermilk or yogurt). Cover pan and leave for 30 minutes at same temperature. Dissolve ¼ rennet tablet in ¼ cup cooled, sterilized water and add to milk, (or put 5 drops of liquid rennet in 1 tablespoon water) per quart. Stir for one minute. Cover pan again and leave undisturbed at the same temperature for one hour until curd is set.

Cut curd carefully into ½ inch cubes and let rest for 10 minutes. Slowly raise temperature to 49°C (120°F) over the period of one hour. Gently stir curds with stainless steel spoon (or clean hands), bringing curds up from the bottom and carefully cutting any that are over ½ inch cubes. Curds will gradually lose their cube shape as they release whey. When they finally look like very large-curd cottage cheese, ladle them into a sterilized cheese cloth placed over a strainer in a large bowl.

Gather corners of cloth and hang cheese to drain over a bowl to catch the whey for an hour or two. Empty curds from cloth into a clean dry bowl and break up into small clumps, working about 2 teaspoons of coarse salt into the curds.

Place salted curds into the sterilized brousse mold, (basket type mold) pressing them down firmly with stainless steel spoon or ladle. Cover top of cheese with a circle of cheese cloth and set mold on an overturned bowl set in a larger bowl to catch the whey. After about an hour, fill a quart jar with water and place on top of cheese to press it.

In about two hours, gently remove cheese from mold and carefully replace it in the mold - upside down. Replace weight and continue pressing process. Turn cheese every few hours until it has taken on the neat, symmetrical shape of the mold, wipe dry with a clean cloth and place on a clean folded towel at room temperature to dry and form a rind. Turn occasionally to allow even drying. After 24 hours, cheese may be placed in cool place for three to five days to age before cutting.

HOW TO MAKE YOUR OWN RENNET

Rennet is an extract made from the lining of the stomach of a ruminant animal. This is used in cheesemaking. You need rennet to make hard cheese.

Method One

Go to the slaughter house and ask for the fourth stomach compartment (Abomasum) of a calf or young goat. Clean and salt the stomach and hang it up to dry in a cool place. Beware of dust and insects. The day before making cheese, break off a small piece of the dried stomach (about the size of your hand). Put in a small container and cover with cooled, sterilized water. Leave for at least three hours. Add ¼ cup of this solution to the milk when you are ready to form the curd. Keep the leftover solution in a cool place. Use within three days.

Note: You can sterilize water by boiling it for 15 minutes and cooling.

Method Two

If a calf or a kid goat is slaughtered or dies and is not more than two days old, the stomach will contain colostrum. Take the colostrum out and save it. Clean the stomach carefully. Return the colostrum to the cleaned stomach. Tie off the opening. Hang in a cool place to age. When the contents are like butter or margarine, it is ready to use. Keep as cool as possible. Mix the colostrum in ¼ cup cool water. Use ½ teaspoon to set 8 liters (2 gallons) of milk

Method Three

Use abomasa from kids that have been fed exclusively on milk (i.e., slaughtered when roughly 1 month old). On receiving the abomasa, clean the outside portions, remove excess fat, close one of the two openings on the abomasum with a well-tied knot. Insert a pipette or straw into the other end and blow up like a balloon. Knot the end to keep the abomasum inflated. Dry in a dry, well-ventilated area for two to three days. Once the abomasa have dried, they keep for several months.

- 2 abomasa, dried as explained above.
- Cut the abomasa into 5 cm (¼ inch) wide strips.
- Add 500 mL water, 45 mL coarse salt, a pinch of boric acid.
- Let steep for 5 days at roughly 12°C (55°F). Stir two or three times during this period. Leave covered.
- Add 250 mL water and 15 mL coarse salt.
- Pour first through an ordinary strainer, then through cheesecloth.
- Next pour through three cheesecloths. Repeat the last operation.

Yogurt

Heat one liter (4 cups) of milk to just below the boiling point for one minute. This destroys bacteria. Cool the milk to just above body temperature. The milk should feel warm, but not hot, on the inside of your wrist.

Put two tablespoons of plain fresh yogurt or dry yogurt into the lukewarm milk. Be sure these are generous spoonfuls. Stir gently, but well. Mix thoroughly with milk. Cover or insulate the pan or pour the inoculated milk into a clean jar or jars. Put on covers. Put containers in a large pot. Put a tight cover on the pot. Wrap pot with heavy towel or blanket and set aside away from drafts or place the pot in the sun. Leave four to eight hours. When yogurt is ready, put in a cool place.

MEAT

Principles and Practices Cooking Goat Meal (Chevon)

Having very low fat content, chevon, when exposed to dry and high temperature cooking environments, will lose moisture and toughen quickly. For best tenderness, juiciness and flavor it is only necessary to observe two basic rules: cook it slowly and baste constantly especially over an open or charcoal fire.

ROSALEE SINN

Goat Meat Chili

- 2 tbs cooking oil
- 2 cups chopped onions
- 1 tbs ground oregano
- 2 tbs ground cumin
- 1 tsp garlic powder 1 tbs salt
- 1.3 kg (3 lb) lean ground goat meat
- ½ cup plus 2 tbs chili powder
- ½ cup flour 1 large can diced tomatoes
- 8 cups boiling water

In heavy pot, sauté onions in cooking oil, add oregano, cumin, garlic powder and salt. Stir and sauté until the onion is almost clear, then add ground meat and cook and stir until crumbly and almost gray. Add chili powder and then the flour, stirring vigorously until thoroughly blended. Lastly, add boiling water and tomatoes and bring mixture to a boil, and simmer for not more than one hour. Seasonings, including cayenne pepper, may be adjusted to individual taste at this time. This recipe makes approximately 14 cups of chili. Adding pinto beans to this chili, before or after cooking, is not recommended; serve the beans as a side dish.

Mexican Rice with Goat Meat

Prepare 1 cup of uncooked rice according to the package directions. While the rice cooks, prepare the following ingredients:

- ½ kg (1 lb) lean ground goat meat
- 1 medium bell pepper, seeded and chopped
- 1 medium onion, peeled and chopped
- 1 can (8 oz) tomato sauce
- 1 heaping tbs chili powder
- ½ tsp salt
- ¼ tsp oregano
- ½ tsp cumin (powdered or seed)

Sauté onion and pepper in 1 tbs. oil; then add ground meat and cook until nearly done, stirring and breaking up with a wooden spoon. Add spices, mix well and then add tomato sauce, stir vigorously. Add drained, cooked rice, mix well. Let stand 15 minutes and serve.

Enchilada Casserole with Goat Meat

- 1 kg (2 lb) lean ground goat meat
- 1 large onion, chopped
- 1 (4 oz) can green chilies, chopped or 3 large chilies, seeded and chopped
- 1 can cream of chicken soup
- 1 can of mushroom soup
- 1 can hot enchilada sauce
- 12 corn tortillas (approximately 5-6" diameter)
- 0.22 kg (½ lb) mild cheddar cheese, grated

Sauté onions in 2 tbs. oil in large skillet. Add meat and brown, breaking up with a spoon. Add chilies, soups and enchilada sauce, mix well, and cook until heated thoroughly. Cut each tortilla in 8 pieces and arrange half in layer in bottom of 13 x 9 x 2 inch baking dish. Cover with layer of meat mixture. Sprinkle half of grated cheese on top of meat. Repeat with second layer. Bake at 350 for 35-45 minutes to yield eight hearty servings.

Curried Chevon

- ½ kg 1 lb Chevon
- 85 gm (3 oz) butter
- 2 tbs minced onion
- 2 tbs fine cut celery
- 2 tbs diced apples
- 1 tbs flour
- 1 tbs curry powder
- 2 ripe tomatoes, stewed and strained
- 1½ cups water
- Salt to taste

Cut meat into one-inch squares, salt and sauté in the butter. Add onion, celery and apples; sauté thoroughly. Sprinkle mixture with flour and curry powder and cook until flour colors. Add strained tomatoes and water, cover saucepan and let cook slowly until done. Serve with steamed rice.

Goat Pot Roast

- 2.26 kg (5 lb) goat shoulder
- 2 cups water
- 1 large onion
- 3 cloves garlic
- Worcestershire sauce
- Salt
- Pepper
- 5 medium potatoes

Put goat meat into roasting pot with water. Sprinkle well with salt, pepper and worcestershire sauce. Add chopped onion and garlic. Put on lowest heat on stove. Cook for five hours. Add potatoes ½ hour before serving.

Goat Meat Curry

- 1.36 kg (3 lb) goat meat
- 2 cardamom
- 2-3 cloves
- 2-3 cinnamon sticks
- 3-4 bay leaves
- 1 tsp whole black pepper
- ¼ cup oil
- 4 chopped onions
- 2 chopped tomatoes
- 2 tbs tomato puree
- 1 garlic paste
- 1 tbs ginger paste
- 2 tbs chopped fresh coriander leaves
- 1 tbs red chili powder
- 1 tbs coriander powder
- 1 tsp turmeric powder
- 1 tbs garam masala
- Salt to taste
- Water for gravy (curry)

Heat oil in frying pan, add cardamom, cloves, cinnamon stick, bay leaves, whole black pepper and fry for few seconds. Then add onions and fry until light brown, add ginger garlic paste, tomato, tomato puree, coriander powder, red chili, turmeric and salt to taste. When masala is thoroughly fried and oil comes up add chevon pieces and fry until brown. Then add water cover pan and keep it on low flame until chevon is done. Garnish with chopped coriander (cilantro) leaves and garam masala for a delicious flavor. Serve with roti or naan (bread).

GOAT MEAT SAUSAGE MAKING

Consumers eat sausages because of convenience, variety, economy, and nutritional value. Goat owners wishing to make sausages from their surplus animals can have an animal slaughtered and boned with the resulting muscle and usable fat in coarse ground or roughly cubed form; other owners may undertake some or all of these chores themselves. Sausage-making procedures for making "kitchen-size" lots of goat meat sausage, one needs only modest equipment, a hand grinder, stuffer, knives and mixing pan. Condiments, cures (chemical powders) and cultures are used in very small amounts and may be purchased as needed or bought in larger lots and stored indefinitely under correct conditions. A smoker device is optional and may be purchased or homemade.

Three requirements are important for successful sausage-making:

- The quality of the original ingredients will dictate in large part the finished product quality,
- Proper sanitation is crucial to avoid off-flavors and possible toxicity, This includes clean hands or wearing clean disposable rubber gloves.
- The meat to be ground and cased should be very cold and handled as quickly and carefully as possible during the processing steps.

Grinding

The typical hand grinder used in this step has three major parts: screw, knife and plate. The diameter of the holes in the plate and the thickness of the plate determine the size of the sausage particles. Certain sausages traditionally are coarse ground, others medium to finely ground. Generally, the smaller the particle size the easier it is to get a homogeneous mass during the subsequent mixing. Fat and lean may be ground simultaneously to promote mixing. If the meat has considerable connective tissue, it will likely be necessary to "clean" the knife and plate at intervals during grinding.

Mixing

Thorough mixing of the ground meat is necessary to get a good blend of lean and fat and to evenly distribute the condiments and cure; improper mixing will appreciably decrease final product quality. Mixing may be accomplished by hand or by using appropriate kitchen mixers. Prolonged mixing is good for uniformity but bad for maintaining cold temperature; as always, compromise is necessary. For some sausages, the maker may elect to first coarse grind (½ inch or more) the meat, then mix, then re-grind more finely (¼ inch or less); homogeneity is thus considerably improved.

Stuffing

The typical hand stuffer used in this step consists of: barrel, piston (plunger) and horn. For larger batches, a screw-type press is used; however, it will have the same components. Obviously, the size and shape of the horn used determines the dimensions of the products extruded. Again, practicality and/or tradition dictates the appropriate horn size for various sausages.

Sausage mixes are stuffed into casings of many types and sizes. Earlier, animal intestines were the primary casing, but now cellulosic casings are the major ones. Four basic types of casings may be purchased: animal, regenerated collagen, cloth and cellulosic; each has advantages and disadvantages and each has product-specific uses. For the home sausage-maker, the synthetic collagen-type, though expensive, may be most desirable. Sausage may also be formed into patties without casings.

Curing

Meat curing was used early on as a means of preserving a fresh product indefinitely. With the advent of home refrigeration, curing of meat is much less necessary; today cured meat products are generally mild cured and thus do require refrigeration.

Salt is basic to all curing mixtures and is actually the only ingredient necessary for curing. Salt alone, however, gives a harsh, dry, salty product that is not adequately palatable and one that produces undesirable color in the lean. Contrarily, salt has beneficial drying and binding properties which provide desirable product firmness. Percentages of salt vary widely by product with 1½ percent to 2 percent being common. Lard is added primarily for flavor, but also for its softening effect and to promote a desirable browning effect during cooking. Sugar is commonly added to sausage mixes at the rate of 0.25 to 2 percent of weight depending on the product.

Perhaps the most important thing to consider in using goat meat for sausage-making is that its typically low fat content may require the use of additional fat to achieve the desired organoleptic characteristics of flavor, juiciness and aroma as well as texture and body. Too little fat does not permit proper processing and may result in reduced product acceptability. The "added fat" may be goat fat or it may be pork or beef fat-trim; pork shoulders and jowls from heavy sows are also popular sources of fat for sausage-making. It is suggested that "homemade" sausage should range from 20 to 25 percent fat. A ratio of 1.3 kg (3 lb) lean goat meat to ½ kg (1 lb) of pork or goat fat would normally achieve this range; alternatively, 1.3 kg (3 lb) of lean goat meat and 0.90 kg (2 lb) of pork shoulder would be satisfactory.

Easy Recipe for Goat Sausage

- 1 lb ground chevon
- 1 tsp sage
- ¼ tsp black pepper
- ½ tsp red pepper
- Salt to taste

Mix well and make into patties. Fry until brown. Do not over-cook. Serve with eggs and potatoes.

REFERENCES

PUBLICATIONS

American Sheep Industry Association. (2004). Special Issue: Predation. *Sheep and Goat Research Journal*, 19 [Special issue].

Bayer, W., & Waters-Bayer, A. (1998). *Forage Husbandry*. London: MacMillan.

Belanger, J. D. (1975). *Raising Milk Goats the Modern Way*. Charlotte, VT: Garden Way Pub. Co.

Burrows, G. E., and Tyrl, R. J. (2001). *Toxic Plants of North America*. Ames: Iowa State University Press.

Campbell, L., (Ed.). (1981). *The Whole Goat Catalog*. Louisa, VA.: Gerard Publishing.

Devendra, C., Burns, M. (1983). *Goat Production in the Tropics*. Farnham Royal: Commonwealth Agricultural Bureaux.

Charlton, J. F. L. (2003). *Using Trees on Farms*. Wellington, N. Z.: New Zealand Grassland Association & New Zealand Farm Forestry Association.

Church, D. C. (1996). *Basic Animal Nutrition & Feeding*. New York: Wiley & Sons.

Dunn, P. (1982). *The Goatkeepers Veterinary Book*. Suffolk, England: Farming Press Limited.

Gall, C., (Ed.). (1981). *Goat Production*. New York: Academic Press.

Gall, C. (1996). *Goat Breeds of the World*. Weikersheim, Germany: Margraf.

Garrett, P. D. (1985-1983). *Guide to Dissection of the Goat*. Opelika, AL: The Author (Rt. 3, Box 309, Opelika, AL 36801)

Grace, N.D. (1983). *The Mineral Requirements of Grazing Ruminants*. Hamilton, NZ: New Zealand Society of Animal Production.

Grandin, T. (1993). *Livestock Handling and Transport*. Wallingford, UK: CAB International.

Haenlein, G. & D. L. Ace (1984-). *Extension Goat Handbook*. Washington D.C.: Extension Service, U. S. Dept. of Agriculture.

Humphreys, L. R. (1978). *Tropical Pastures and Fodder Crops*. London: Longman Group.

Luttman, G. (1986). *Raising Milk Goats Successfully*. Charlotte, VT: Williamson Publishing.

Hill, H. J. (1985). *The International Goat Gourmet*. Olathe, KS: Cookbook Publishers.

MacKenzie, D. (1980). *Goat Husbandry*. London: Faber & Faber.

Matthews, J. G. (1999). *Diseases of the Goat*. Ames: Iowa State University Press.

McBrier, P. & L. Lohstoeter. (2001). *Beatrice's Goat*, New York: Atheneum Books for Young Readers.

McDowell, L. R. (1985) *Nutrition of Grazing Ruminants in Warm Climates*. Orlando, FL: Academic Press.

Mollison, B. (1988). *Permaculture – A Designers' Manual*. Tyalgum, Australia: Tagari Publications.

National Research Council (U.S.). (2007). *Nutrient Requirements of Small Ruminants*. Washington D. C.: The National Academies Press.

Naude, R. T. and H. S. Hofmeyer. (1981). *Meat Production in Goat Production*. New York: Academic Press.

Nelson-Henrich, S. (1980). *The Complete Book of Yogurt*. New York: Macmillan.

Park, Y. W., Haenlein, G. F.W. (2006). *Handbook of Milk of Non-Bovine Mammals*. New York: Blackwell Publishing.

Peischel, A. (Ed.). (2007). *Master Meat Goat Producer*. Nashville: Cooperative Extension Program, Tennessee State University.

Peacock, C. P. (1996). *Improving Goat Production in the Tropics*. Oxford: Oxfam in association with FARM-Africa.

Pinkerton, F., Scarfe, D. & B. Pinkerton (1991). *Meat Goat Production and Marketing No. M-01*. E. (Kika) de la Garza Institute for Goat Research, Langston University. Retrieved September 7, 2007 from www2.luresext.edu/goats/library/fact_sheets/m01.htm

Plucknett, D. (1979). *Managing Pastures and Cattle under Coconuts*. Boulder, CO: Westview Press.

Pugh, D.G. (2002). *Sheep and Goat Medicine*. Philadelphia: Saunders.

Porter, V. (1996). *Goats of the World*. Ipswich, UK: Farming Press.

Rubino, R., Morand-Fehr, P. & L. Sepe. (2004). *Atlas of Goat Products*. Potenza, Italy: La biblioteca di Caseus.

Savory, A. (1998). *Holistic Resource Management Washington*. D.C.: Island Press.

Sivaraj, S., Agamuthu, P. & Mukherjee, T.K. (1992) *Advances in Sustainable Small Ruminant-Tree Cropping Integrated Systems*. Kuala Lumpur: Organising Secretariat IPT/ IDRC (International Development Research Centre) Workshop on Development of Sustainable Integrated Small Ruminant-Tree Cropping Production Systems.

Smith, M. & D. Sherman. (1994). *Goat Medicine*. Philadelphia, PA: Lea & Febiger.

Stanton, T. (n.d.) *Goat Management*. Animal Science Department, Cornell University. Retrieved on September 7, 2007 from www.ansci.cornell.edu/goats/resources_list.html

Stewart. S. (1998). *Learning Together, The Agricultural Worker's Participatory Sourcebook*. Little Rock, AR: Heifer International; Seattle WA: Christian Veterinary Mission.

Vanderhoof, R. A., (1990). *Raising Healthy Goats Under Primitive Conditions*. Seattle, WA: Christian Veterinary Mission.

Vella, Jane. (1995). *Training Through Dialogue: Promoting Effective Learning and Change with Adults*. Jossy-Bass Inc. San Francisco, CA.

Yerex, D. (1986). *The Farming of Goats*. Carterton, New Zealand: Ampersand Publishing Association.

INTERNET SITES

GOAT HUSBANDRY - GENERAL

ATTRA - National Sustainable Agriculture Information Service List of Small Ruminant Resources
http://attra.ncat.org/attra-pub/small_ruminant_resources.html

Breeds: Oklahoma State University Livestock Breeds Online Guide
www.ansi.okstate.edu/breeds/goats

Bucks, Their Breeding Habits
Fias Co. Farm Guide To Bucks and Breeding Habits
www.fiascofarm.com/goats/buck-wether-info.htm

Equipment and Supplies: Caprine Supply
www.caprinesupply.com

Guide to Goat Management, Cornell University Animal Science Department
www.ansci.cornell.edu/goats

Maryland Small Ruminant Page
www.sheepandgoat.com

Raising Goats for Milk and Meat
For more information on this book: www.heifer.org/programinfo

Research and Extension on Goats, Langston University
www2.luresext.edu/goats/research/currentresearchprojects.htm

GOAT HUSBANDRY - DAIRY GOATS AND MILK PRODUCTS

Cheese Making: New England Cheesemaking Supply Company
www.cheesemaking.com
info@cheesemaking.com

Dairy Goat Guide, Washington State University
cru.cahe.wsu.edu/CEPublications/em4894/em4894.pdf

Dairy Goat Milk Composition
www.goatworld.com/articles/goatmilkcomposition.shtml

Guide to Milk and Dairy Products, UN FAO Animal Production and Health Division
www.fao.org/ag/againfo/subjects/en/dairy/home.html

Understanding Dairy Goat Production
www.goatworld.com/articles/udgp.shtml

GOAT HUSBANDRY – MEAT GOATS

Agricultural Extension Service, University of Tennessee
www.utextension.utk.edu

Body Condition Scoring
www.cals.ncsu.edu/an_sci/extension/animal/meatgoat/MG-BCS.htm

Meat Goat Production and Marketing Handbook, Clemson University
www.clemson.edu/agrononomy/goats/handbook/toc.html

Meat Goats, North Carolina State University College of Agriculture and Life Sciences
www.cals.ncsu.edu/an_sci/extension/animal/meatgoat/ahgoats_index.html

GOAT HEALTH

Animal Health Resources: UN FAO Animal Production and Health Division
www.fao.org/ag/againfo/subjects/en/health/default.html

Biology of the Goat
www.imagecyte.com/goats.html
Ethonoveterinary Web
www.ethnovetweb.com

Parasites, (Including Information on Famacha):
Southern Consortium for Small Ruminant Parasite Control
www.scsrpc.org
famacha@vet.uga.edu

Poisonous Plants: Cornell Poisonous Plants Informational Database
www.ansci.cornell.edu/plants/alphalist.html

Primary Animal Health Care Workers, FAO Document Repository
www.fao.org/documents (Search under "animal health")
www.fao.org/docrep/T0690E/T0690E00.htm

NUTRITION AND FORAGE PRODUCTION

Cassava - Daleys Fruit Tree Nursery Information and Product Catalog
www.daleysfruit.com.au/fruit%20pages/cassava.htm

The Overstory International Agroforestry E-journal
www.agroforestry.net/overstory/osprev.html

Tropical Forages: An Interactive Selection Tool
www.tropicalforages.info

GOAT

Small Creature
You enter the world
And stand up to face
Opportunity
Suckling, jumping
Exploring Life

Small Companion
You join our family
Almost like laughter
Joyfully
Animal friend
Pursuing life

Small Gatherer
You consume the bush
Climb the mountains
Foraging
Standing tall
Protecting life

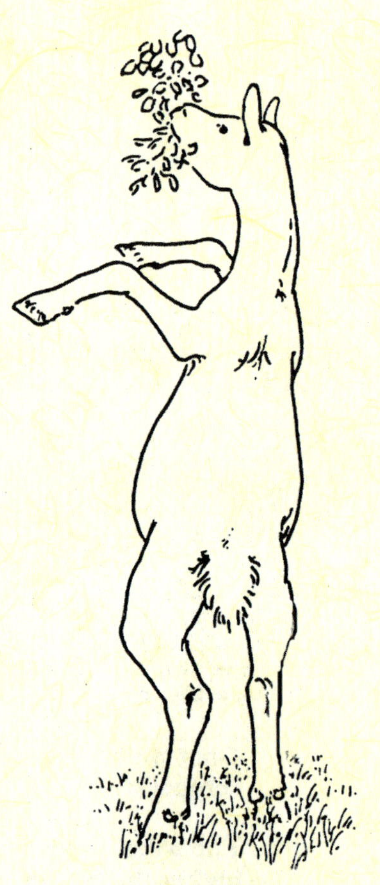

Small Provider
You give milk and meat
Sharing good health
Growing
Giving food
Nourishing life

Small Wonder
You bear the burden
Of ignorance and pride
Focusing
Always outward
Sharing life

Small Goat
You grow in stature
Your gifts unequalled
Seeking
Asking a chance
To feed the world!

—Rosalee Sinn

NOTES

NOTES